The Positron

" The Anti-particle of The Electron "

Edited by Paul F. Kisak

Contents

Chapter 1

Positron

For other uses, see Positron (disambiguation).

The **positron** or **antielectron** is the antiparticle or the antimatter counterpart of the electron. The positron has an electric charge of +1 e, a spin of ½, and has the same mass as an electron. When a low-energy positron collides with a low-energy electron, annihilation occurs, resulting in the production of two or more gamma ray photons (see electron–positron annihilation).

Positrons may be generated by positron emission radioactive decay (through weak interactions), or by pair production from a sufficiently energetic photon which is interacting with an atom in a material.

1.1   History

1.1.1   Theory

In 1928, Paul Dirac published a paper[2] proposing that electrons can have both a positive charge and negative energy. This paper introduced the Dirac equation, a unification of quantum mechanics, special relativity, and the then-new concept of electron spin to explain the Zeeman effect. The paper did not explicitly predict a new particle, but did allow for electrons having either positive or negative energy as solutions. Hermann Weyl then published "Gravitation and the Electron" (Proceedings of the National Academy of Sciences of the United States of America, Vol. 15, No. 4-Apr. 15, 1929, pp. 323–334) discussing the mathematical implications of the negative energy solution. The positive-energy solution explained experimental results, but Dirac was puzzled by the equally valid negative-energy solution that the mathematical model allowed. Quantum mechanics did not allow the negative energy solution to simply be ignored, as classical mechanics often did in such equations; the dual solution implied the possibility of an electron spontaneously jumping between positive and negative energy states. However, no such transition had yet been observed experimentally. He referred to the issues raised by this conflict between theory and observation as "difficulties" that were "unresolved".

Dirac wrote a follow-up paper in December 1929[3] that attempted to explain the unavoidable negative-energy solution for the relativistic electron. He argued that "... an electron with negative energy moves in an external [electromagnetic] field as though it carries a positive charge." He further asserted that all of space could be regarded as a "sea" of negative energy states that were filled, so as to prevent electrons jumping between positive energy states (negative electric charge) and negative energy states (positive charge). The paper also explored the possibility of the proton being an island in this sea, and that it might actually be a negative-energy electron. Dirac acknowledged that the proton having a much greater mass than the electron was a problem, but expressed "hope" that a future theory would resolve the issue.

Robert Oppenheimer argued strongly against the proton being the negative-energy electron solution to Dirac's equation. He asserted that if it were, the hydrogen atom would rapidly self-destruct.[4] Persuaded by Oppenheimer's argument, Dirac published a paper in 1931 that predicted the existence of an as-yet unobserved particle that he called an "anti-electron" that would have the same mass as an electron and that would mutually annihilate upon contact with an electron.[5]

Feynman, and earlier Stueckelberg, proposed an interpretation of the positron as an electron moving backward in time,[6] reinterpreting the negative-energy solutions of the Dirac equation. Electrons moving backward in time would have a positive electric charge. Wheeler invoked this concept to explain the identical properties shared by all electrons, suggesting that "they are all the same electron" with a complex, self-intersecting worldline.[7] Yoichiro Nambu later applied it to all production and annihilation of particle-antiparticle pairs, stating that "the eventual creation and annihilation of pairs that may occur now and then is no creation or annihilation, but only a change of direction of moving particles, from past to future, or from future to past."[8] The backwards in time point of view is nowadays accepted as completely equivalent to other pictures, but it does not have anything to do with the macroscopic terms "cause" and "effect", which do not appear in a microscopic physical description.

1.1.2   Experimental clues and discovery

Dmitri Skobeltsyn first observed the positron in 1929.[9][10] While using a Wilson cloud chamber[11] to try to detect gamma radiation in cosmic rays, Skobeltsyn detected particles that acted like electrons but curved in the opposite direction in an applied magnetic field.[10]

Likewise, in 1929 Chung-Yao Chao, a graduate student at Caltech, noticed some anomalous results that indicated particles behaving like electrons, but with a positive charge, though the results were inconclusive and the phenomenon was not pursued.[12]

Carl David Anderson discovered the positron on August 2, 1932,[13] for which he won the Nobel Prize for Physics in 1936.[14] Anderson did not coin the term *positron*, but allowed it at the suggestion of the Physical Review journal editor to which he submitted his discovery paper in late 1932. The positron was the first evidence of antimatter and was discovered when Anderson allowed cosmic rays to pass through a cloud chamber and a lead plate. A magnet surrounded this apparatus, causing particles to bend in different directions based on their electric charge. The ion trail left by each positron appeared on the photographic plate with a curvature matching the mass-to-charge ratio of an electron, but in a direction that showed its charge was positive.[15]

Anderson wrote in retrospect that the positron could have been discovered earlier based on Chung-Yao Chao's work, if only it had been followed up.[12] Frédéric and Irène Joliot-Curie in Paris had evidence of positrons in old photographs when Anderson's results came out, but they had dismissed them as protons.[15]

1.2   Natural production

Main article: Positron emission

Positrons are produced naturally in β^+ decays of naturally occurring radioactive isotopes (for example, potassium-40) and in interactions of gamma quanta (emitted by radioactive nuclei) with matter. Antineutrinos are another kind of antiparticle created by natural radioactivity (β^- decay). Many different kinds of antiparticles are also produced by (and contained in) cosmic rays. Recent (as of January 2011) research by the American Astronomical Society has discovered antimatter (positrons) originating above thunderstorm clouds; positrons are produced in gamma-ray flashes created by electrons accelerated by strong electric fields in the clouds.[16] Antiprotons have also been found to exist in the Van Allen Belts around the Earth by the PAMELA module.[17][18]

Antiparticles, of which the most common are positrons due to their low mass, are also produced in any environment with a sufficiently high temperature (mean particle energy greater than the pair production threshold). During the period of baryogenesis, when the universe was extremely hot and dense, matter and antimatter were continually produced and annihilated. The presence of remaining matter, and absence of detectable remaining antimatter,[19] also called baryon asymmetry, is attributed to CP-violation: a violation of the CP-symmetry relating matter to antimatter. The exact mechanism of this violation during baryogenesis remains a mystery.

Positrons production from radioactive β+ decay, can be considered both artificial and natural production, as the generation of the radioisotope can be natural or artificial. Perhaps the best known naturally-occurring radioisotope which produces positrons is potassium-40, a long-lived isotope of potassium which occurs as a primordial isotope of potassium, and even though a small percent of potassium, (0.0117%) is the single most abundant radioisotope in the human body. In a human body of 70 kg mass, about 4,400 nuclei of ^{40}K decay per second.[20] The activity of natural potassium is 31 Bq/g.[21] About 0.001% of these ^{40}K decays produce about 4000 natural positrons per day in the human body.[22] These positrons soon find an electron, undergo annihilation, and produce pairs of 511 keV gamma rays, in a process similar (but much lower intensity) to that which happens during a PET scan nuclear medicine procedure.

1.2.1 Observation in cosmic rays

Main article: Cosmic ray

Satellite experiments have found evidence of positrons (as well as a few antiprotons) in primary cosmic rays, amounting to less than 1% of the particles in primary cosmic rays. These do not appear to be the products of large amounts of antimatter from the Big Bang, or indeed complex antimatter in the universe (evidence for which is lacking, see below). Rather, the antimatter in cosmic rays appear to consist of only these two elementary particles, probably made in energetic processes long after the Big Bang.

Preliminary results from the presently operating Alpha Magnetic Spectrometer (*AMS-02*) on board the International Space Station show that positrons in the cosmic rays arrive with no directionality, and with energies that range from 10 GeV to 250 GeV. In September, 2014, new results with almost twice as much data were presented in a talk at CERN and published in Physical Review Letters.[23][24] A new measurement of positron fraction up to 500 GeV was reported, showing that positron fraction peaks at a maximum of about 16% of total electron+positron events, around an energy of 275 ± 32 GeV. At higher energies, up to 500 GeV, the ratio of positrons to electrons begins to fall again. The absolute flux of positrons also begins to fall before 500 GeV, but peaks at energies far higher than electron energies, which peak about 10 GeV.[25] These results on interpretation have been suggested to be due to positron production in annihilation events of massive dark matter particles.[26]

Positrons, like anti-protons, do not appear to originate from any hypothetical "antimatter" regions of the universe. On the contrary, there is no evidence of complex antimatter atomic nuclei, such as antihelium nuclei (i.e., anti-alpha particles), in cosmic rays. These are actively being searched for. A prototype of the *AMS-02* designated *AMS-01*, was flown into space aboard the Space Shuttle *Discovery* on STS-91 in June 1998. By not detecting any antihelium at all, the *AMS-01* established an upper limit of 1.1×10^{-6} for the antihelium to helium flux ratio.[27]

1.3 Artificial production

New research has dramatically increased the quantity of positrons that experimentalists can produce. Physicists at the Lawrence Livermore National Laboratory in California have used a short, ultra-intense laser to irradiate a millimetre-thick gold target and produce more than 100 billion positrons.[28][29]

1.4 Applications

Certain kinds of particle accelerator experiments involve colliding positrons and electrons at relativistic speeds. The high impact energy and the mutual annihilation of these matter/antimatter opposites create a fountain of diverse subatomic particles. Physicists study the results of these collisions to test theoretical predictions and to search for new kinds of particles.

Gamma rays, emitted indirectly by a positron-emitting radionuclide (tracer), are detected in positron emission tomography (PET) scanners used in hospitals. PET scanners create detailed three-dimensional images of metabolic activity within the human body.[30]

An experimental tool called positron annihilation spectroscopy (PAS) is used in materials research to detect variations in density, defects, displacements, or even voids, within a solid material.[31]

1.5   See also

- Beta particle
- Radioactive decay
- List of particles
- Positron emission tomography
- Positronium
- Proton
- Positronic brain

1.6   References

1.6.1   Notes

[1] The fractional version's denominator is the inverse of the decimal value (along with its relative standard uncertainty of 4.2×10^{-10}).

1.6.2   Citations

[1] The original source for CODATA is:

Mohr, P.J.; Taylor, B.N.; Newell, D.B. (2006). "CODATA recommended values of the fundamental physical constants". *Reviews of Modern Physics* **80** (2): 633–730. arXiv:0801.0028. Bibcode:2008RvMP...80..633M. doi:10.1103/RevModPhys.80.633.
Individual physical constants from the CODATA are available at:

"The NIST Reference on Constants, Units and Uncertainty". National Institute of Standards and Technology. Retrieved 2013-10-24.

[2] P. A. M. Dirac. "The quantum theory of the electron" (PDF).

[3] P. A. M. Dirac. "A Theory of Electrons and Protons" (PDF).

[4] Frank Close (2009). *Antimatter*. Oxford University Press. p. 46. ISBN 978-0-19-955016-6.

[5] P. A. M. Dirac (1931). "Quantised Singularities in the Quantum Field". *Proc. R. Soc. Lond.* A**133**(821): 2–3. Bibcode: doi:10.1098/rspa.1931.0130.

[6] Feynman, Richard (1949). "The Theory of Positrons". *Physical Review* **76**(76): 749. Bibcode:1949PhRv...76..749F.

[7] Feynman, Richard (1965-12-11). *The Development of the Space-Time View of Quantum Electrodynamics* (Speech). Lecture. Retrieved 2007-01-02.

[8] Nambu, Yoichiro (1950). "The Use of the Proper Time in Quantum Electrodynamics I". *Progress in Theoretical Physics* 82. Bibcode:1950PThPh...5...82N. doi:10.1143/PTP.5.82.

[9] Frank Close. *Antimatter*. Oxford University Press. pp. 50–52. ISBN 978-0-19-955016-6.

[10] *general chemistry*. Taylor & Francis. 1943. p. 660. GGKEY:0PYLHBL5D4L. Retrieved 15 June 2011.

[11] Cowan, Eugene (1982). "The Picture That Was Not Reversed". *Engineering & Science* **46** (2): 6–28.

[12] Jagdish Mehra; Helmut Rechenberg (2000). *The Historical Development of Quantum Theory, Volume 6: The Quantum Mechanics 1926–1941*. Springer. p. 804. ISBN 978-0-387-95175-1.

[13] Anderson, Carl D. (1933). "The Positive Electron". *Physical Review* **43**(6): 491–494. Bibcode:1933PhRv doi:10.1103/PhysRev.43.491.

[14] "The Nobel Prize in Physics 1936". Retrieved 2010-01-21.

[15] GILMER, PENNY J. (19 July 2011). "IRÈNE JOLIOT-CURIE, A NOBEL LAUREATE IN ARTIFICIAL RADIOACTIV-ITY" (PDF). p. 8. Retrieved 13 July 2013.

[16] "Antimatter caught streaming from thunderstorms on Earth". BBC. 11 January 2011. Archived from the original on 12 January 2011. Retrieved 11 January 2011.

[17] Adriani, O.; Barbarino, G. C.; Bazilevskaya, G. A.; Bellotti, R.; et al. (2011). "The Discovery of Geomagnetically Trapped Cosmic-Ray Antiprotons". *The Astrophysical Journal Letters* **737** (2): L29. arXiv:1107.4882v1. Bibcode:2011ApJ...737L..29A. doi:10.1088/2041-8205/737/2/L29.

[18] Than, Ker (10 August 2011). "Antimatter Found Orbiting Earth—A First". National Geographic Society. Retrieved 12 August 2011.

[19] "What's the Matter with Antimatter?". NASA. 29 May 2000. Archived from the original on 4 June 2008. Retrieved 24 May 2008.

[20] "Radiation and Radioactive Decay. Radioactive Human Body". Harvard Natural Sciences Lecture Demonstrations. Retrieved 2011-05-18.

[21] Winteringham, F. P. W; Effects, F.A.O. Standing Committee on Radiation, Land And Water Development Division, Food and Agriculture Organization of the United Nations (1989). *Radioactive fallout in soils, crops and food: a background review*. Food & Agriculture Org. p. 32. ISBN 978-92-5-102877-3.

[22] Engelkemeir, DW; KF Flynn; LE Glendenin (1962). "Positron Emission in the Decay of K^{40}". *Physical Review* **126** (5): 1818. Bibcode:1962PhRv..126.1818E. doi:10.1103/PhysRev.126.1818.

[23] L. Accardo et al. (AMS Collaboration) (18 September 2014). "High Statistics Measurement of the Positron Fraction in Primary Cosmic Rays of 0.5–500 GeV with the Alpha Magnetic Spectrometer on the International Space Station" (PDF). *Physical Review Letters* **113**: 121101. Bibcode:2014PhRvL.113l1101A. doi:10.1103/PhysRevLett.113.121101.

[24] Schirber, Michael. "Synopsis: More Dark Matter Hints from Cosmic Rays?". American Physical Society. Retrieved 21 September 2014.

[25] "New results from the Alpha Magnetic$Spectrometer on the International Space Station" (PDF). *AMS-02 at NASA*. Retrieved 21 September 2014.

[26] Aguilar, M.; Alberti, G.; Alpat, B.; Alvino, A.; Ambrosi, G.; Andeen, K.; Anderhub, H.; Arruda, L.; Azzarello, P.; Bachlechner, A.; Barao, F.; Baret, B.; Barrau, A.; Barrin, L.; Bartoloni, A.; Basara, L.; Basili, A.; Batalha, L.; Bates, J.; Battiston, R.; Bazo, J.; Becker, R.; Becker, U.; Behlmann, M.; Beischer, B.; Berdugo, J.; Berges, P.; Bertucci, B.; Bigongiari, G.; et al. (2013). "First Result from the Alpha Magnetic Spectrometer on the International Space Station: Precision Measurement of the Positron Fraction in Primary Cosmic Rays of 0.5–350 GeV". *Physical Review Letters* **110** (14): 141102. Bibcode:2013PhRvL.110n1102A. doi:10.1103/PhysRevLett.110.141102.

[27] AMS Collaboration; Aguilar, M.; Alcaraz, J.; Allaby, J.; Alpat, B.; Ambrosi, G.; Anderhub, H.; Ao, L.; et al. (August 2002). "The Alpha Magnetic Spectrometer (AMS) on the International Space Station: Part I – results from the test flight on the space shuttle". *Physics Reports* **366** (6): 331–405. Bibcode:2002PhR...366..331A. doi:10.1016/S0370-1573(02)00013-3.

[28] Bland, E. (1 December 2008). "Laser technique produces bevy of antimatter". MSNBC. Retrieved 2009-07-16. The LLNL scientists created the positrons by shooting the lab's high-powered Titan laser onto a one-millimeter-thick piece of gold.

[29] "Laser creates billions of antimatter particles". *Cosmos Online*.

[30] Phelps, Michael E. (2006). *PET: physics, instrumentation, and scanners*. Springer. pp. 2–3. ISBN 0-387-32302-3.

[31] "Introduction to Positron Research". *St. Olaf College*.

1.7   External links

- What is a Positron? (from the Frequently Asked Questions :: Center for Antimatter-Matter Studies)

- Website about positrons and antimatter

- Positron information search at SLAC

- Positron Annihilation as a method of experimental physics used in materials research.

- New production method to produce large quantities of positrons

- Website about antimatter (positrons, positronium and antihydrogen). Positron Laboratory, Como, Italy

- Website of the AEgIS: Antimatter Experiment: Gravity, Interferometry, Spectroscopy, CERN

- Synopsis: Tabletop Particle Accelerator ... new tabletop method for generating electron-positron streams.

Chapter 2

Antiparticle

Corresponding to most kinds of particles, there is an associated antimatter **antiparticle** with the same mass and opposite charge (including electric charge). For example, the antiparticle of the electron is the positively charged positron, which is produced naturally in certain types of radioactive decay.

The laws of nature are very nearly symmetrical with respect to particles and antiparticles. For example, an antiproton and a positron can form an antihydrogen atom, which is believed to have the same properties as a hydrogen atom. This leads to the question of why the formation of matter after the Big Bang resulted in a universe consisting almost entirely of matter, rather than being a half-and-half mixture of matter and antimatter. The discovery of Charge Parity violation helped to shed light on this problem by showing that this symmetry, originally thought to be perfect, was only approximate.

Particle-antiparticle pairs can annihilate each other, producing photons; since the charges of the particle and antiparticle are opposite, total charge is conserved. For example, the positrons produced in natural radioactive decay quickly annihilate themselves with electrons, producing pairs of gamma rays, a process exploited in positron emission tomography.

Antiparticles are produced naturally in beta decay, and in the interaction of cosmic rays in the Earth's atmosphere. Because charge is conserved, it is not possible to create an antiparticle without either destroying a particle of the same charge (as in beta decay) or creating a particle of the opposite charge. The latter is seen in many processes in which both a particle and its antiparticle are created simultaneously, as in particle accelerators. This is the inverse of the particle-antiparticle annihilation process.

Although particles and their antiparticles have opposite charges, electrically neutral particles need not be identical to their antiparticles. The neutron, for example, is made out of quarks, the antineutron from antiquarks, and they are distinguishable from one another because neutrons and antineutrons annihilate each other upon contact. However, other neutral particles are their own antiparticles, such as photons, hypothetical gravitons, and some WIMPs.

2.1 History

2.1.1 Experiment

In 1932, soon after the prediction of positrons by Paul Dirac, Carl D. Anderson found that cosmic-ray collisions produced these particles in a cloud chamber— a particle detector in which moving electrons (or positrons) leave behind trails as they move through the gas. The electric charge-to-mass ratio of a particle can be measured by observing the radius of curling of its cloud-chamber track in a magnetic field. Positrons, because of the direction that their paths curled, were at first mistaken for electrons travelling in the opposite direction. Positron paths in a cloud-chamber trace the same helical path as an electron but rotate in the opposite direction with respect to the magnetic field direction due to their having the same magnitude of charge-to-mass ratio but with opposite charge and, therefore, opposite signed charge-to-mass ratios.

The antiproton and antineutron were found by Emilio Segrè and Owen Chamberlain in 1955 at the University of California, Berkeley. Since then, the antiparticles of many other subatomic particles have been created in particle accelerator

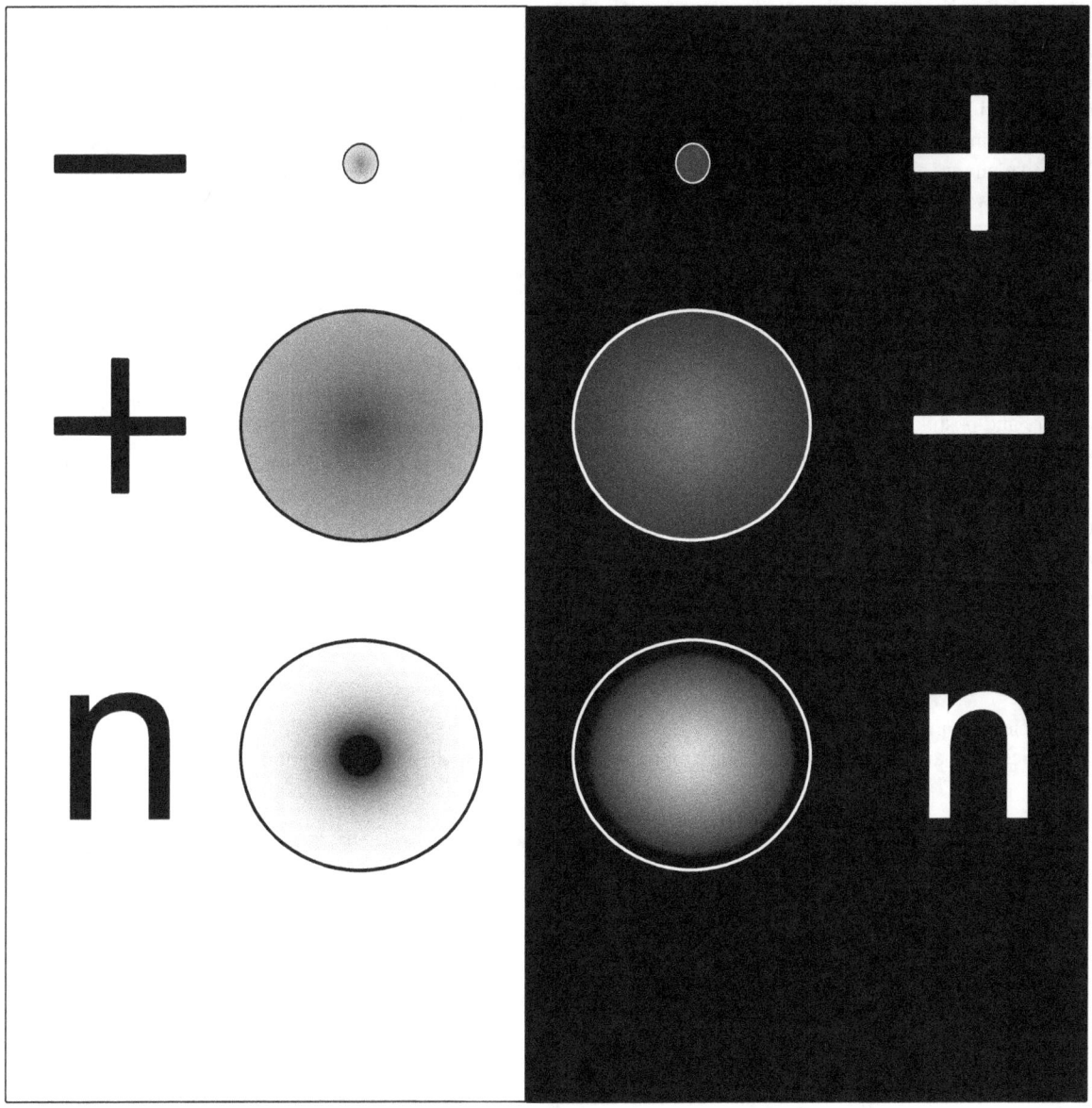

Illustration of electric charge of particles (left) and antiparticles (right). From top to bottom; electron/positron, proton/antiproton, neutron/antineutron.

experiments. In recent years, complete atoms of antimatter have been assembled out of antiprotons and positrons, collected in electromagnetic traps.[1]

2.1.2 Dirac's Hole theory

... the development of quantum field theory made the interpretation of antiparticles as holes unnecessary, even though it lingers on in many textbooks.

Steven Weinberg[2]

Solutions of the Dirac equation contained negative energy quantum states. As a result, an electron could always radiate energy and fall into a negative energy state. Even worse, it could keep radiating infinite amounts of energy because there were infinitely many negative energy states available. To prevent this unphysical situation from happening, Dirac proposed that a "sea" of negative-energy electrons fills the universe, already occupying all of the lower-energy states so that, due to the Pauli exclusion principle, no other electron could fall into them. Sometimes, however, one of these negative-energy particles could be lifted out of this Dirac sea to become a positive-energy particle. But, when lifted out, it would leave behind a *hole* in the sea that would act exactly like a positive-energy electron with a reversed charge. These he interpreted as "negative-energy electrons" and attempted to identify them with protons in his 1930 paper *A Theory of Electrons and Protons*[3] However, these "negative-energy electrons" turned out to be positrons, and not protons.

This picture implied an infinite negative charge for the universe--a problem of which Dirac was aware. Dirac tried to argue that we would perceive this as the normal state of zero charge. Another difficulty was the difference in masses of the electron and the proton. Dirac tried to argue that this was due to the electromagnetic interactions with the sea, until Hermann Weyl proved that hole theory was completely symmetric between negative and positive charges. Dirac also predicted a reaction $e- + p+ \rightarrow \gamma + \gamma$, where an electron and a proton annihilate to give two photons. Robert Oppenheimer and Igor Tamm proved that this would cause ordinary matter to disappear too fast. A year later, in 1931, Dirac modified his theory and postulated the positron, a new particle of the same mass as the electron. The discovery of this particle the next year removed the last two objections to his theory.

However, the problem of infinite charge of the universe remains. Also, as we now know, bosons also have antiparticles, but since bosons do not obey the Pauli exclusion principle (only fermions do), hole theory does not work for them. A unified interpretation of antiparticles is now available in quantum field theory, which solves both these problems.

2.2 Particle-antiparticle annihilation

Main article: Annihilation

If a particle and antiparticle are in the appropriate quantum states, then they can annihilate each other and produce

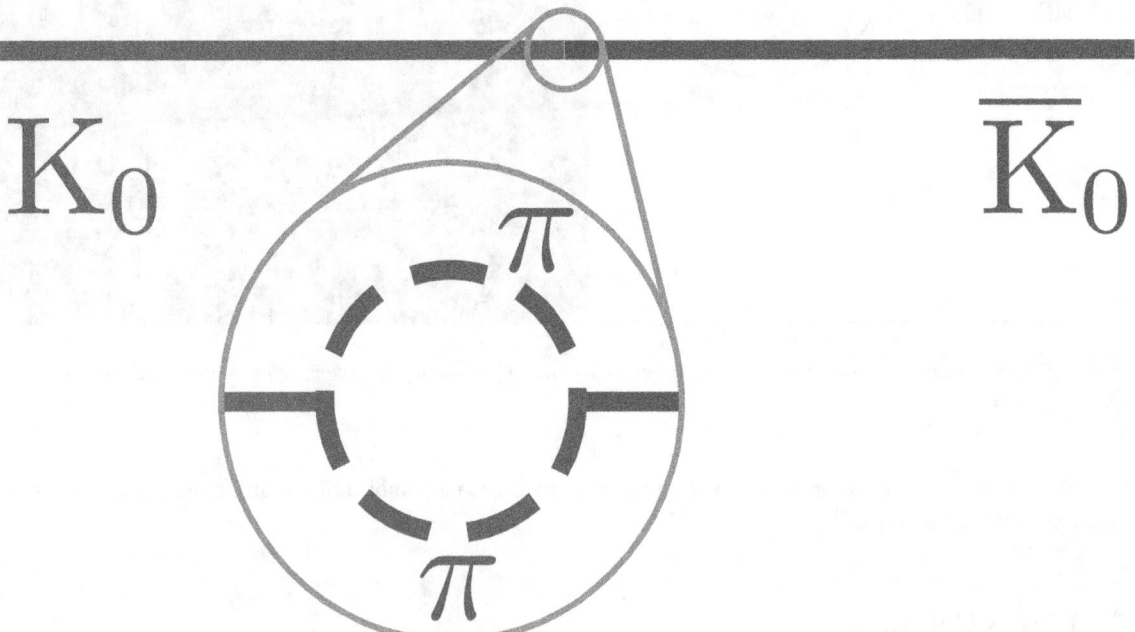

An example of a virtual pion pair that influences the propagation of a kaon, causing a neutral kaon to mix *with the antikaon. This is an example of renormalization in quantum field theory— the field theory being necessary because of the change in particle number.*

other particles. Reactions such as $e- + e+ \rightarrow \gamma + \gamma$ (the two-photon annihilation of an electron-positron pair) are an example. The single-photon annihilation of an electron-positron pair, $e- + e+ \rightarrow \gamma$, cannot occur in free space because it is impossible to conserve energy and momentum together in this process. However, in the Coulomb field of a nucleus the translational invariance is broken and single-photon annihilation may occur.[4] The reverse reaction (in free space, without an atomic nucleus) is also impossible for this reason. In quantum field theory, this process is allowed only as an intermediate quantum state for times short enough that the violation of energy conservation can be accommodated by the uncertainty principle. This opens the way for virtual pair production or annihilation in which a one particle quantum state may *fluctuate* into a two particle state and back. These processes are important in the vacuum state and renormalization of a quantum field theory. It also opens the way for neutral particle mixing through processes such as the one pictured here, which is a complicated example of mass renormalization.

2.3   Properties of antiparticles

Quantum states of a particle and an antiparticle can be interchanged by applying the charge conjugation (**C**), parity (**P**), and time reversal (**T**) operators. If $|p, \sigma, n\rangle$ denotes the quantum state of a particle (**n**) with momentum **p**, spin **J** whose component in the z-direction is σ, then one has

$$CPT \, |p, \sigma, n\rangle \;=\; (-1)^{J-\sigma} \, |p, -\sigma, n^c\rangle,$$

where n^c denotes the charge conjugate state, *i.e.*, the antiparticle. This behaviour under **CPT** is the same as the statement that the particle and its antiparticle lie in the same irreducible representation of the Poincaré group. Properties of antiparticles can be related to those of particles through this. If **T** is a good symmetry of the dynamics, then

$$T \, |p, \sigma, n\rangle \;\propto\; |-p, -\sigma, n\rangle,$$

$$CP \, |p, \sigma, n\rangle \;\propto\; |-p, \sigma, n^c\rangle,$$

$$C \, |p, \sigma, n\rangle \;\propto\; |p, \sigma, n^c\rangle,$$

where the proportionality sign indicates that there might be a phase on the right hand side. In other words, particle and antiparticle must have

- the same mass **m**
- the same spin state **J**
- opposite electric charges **q** and **-q**.

2.4   Quantum field theory

This section draws upon the ideas, language and notation of canonical quantization of a quantum field theory.

One may try to quantize an electron field without mixing the annihilation and creation operators by writing

$$\psi(x) = \sum_k u_k(x) a_k e^{-iE(k)t},$$

where we use the symbol k to denote the quantum numbers p and σ of the previous section and the sign of the energy, $E(k)$, and ak denotes the corresponding annihilation operators. Of course, since we are dealing with fermions, we have to have the operators satisfy canonical anti-commutation relations. However, if one now writes down the Hamiltonian

$$H = \sum_k E(k) a_k^\dagger a_k,$$

then one sees immediately that the expectation value of H need not be positive. This is because $E(k)$ can have any sign whatsoever, and the combination of creation and annihilation operators has expectation value 1 or 0.

So one has to introduce the charge conjugate *antiparticle* field, with its own creation and annihilation operators satisfying the relations

$$b_{k\prime} = a_k^\dagger \text{ and } b_{k\prime}^\dagger = a_k,$$

where k has the same p, and opposite σ and sign of the energy. Then one can rewrite the field in the form

$$\psi(x) = \sum_{k_+} u_k(x) a_k e^{-iE(k)t} + \sum_{k_-} u_k(x) b_k^\dagger e^{-iE(k)t},$$

where the first sum is over positive energy states and the second over those of negative energy. The energy becomes

$$H = \sum_{k_+} E_k a_k^\dagger a_k + \sum_{k_-} |E(k)| b_k^\dagger b_k + E_0,$$

where E_0 is an infinite negative constant. The vacuum state is defined as the state with no particle or antiparticle, *i.e.*, $a_k|0\rangle = 0$ and $b_k|0\rangle = 0$. Then the energy of the vacuum is exactly E_0. Since all energies are measured relative to the vacuum, **H** is positive definite. Analysis of the properties of ak and bk shows that one is the annihilation operator for particles and the other for antiparticles. This is the case of a fermion.

This approach is due to Vladimir Fock, Wendell Furry and Robert Oppenheimer. If one quantizes a real scalar field, then one finds that there is only one kind of annihilation operator; therefore, real scalar fields describe neutral bosons. Since complex scalar fields admit two different kinds of annihilation operators, which are related by conjugation, such fields describe charged bosons.

2.4.1   Feynman–Stueckelberg interpretation

By considering the propagation of the negative energy modes of the electron field backward in time, Ernst Stueckelberg reached a pictorial understanding of the fact that the particle and antiparticle have equal mass **m** and spin **J** but opposite charges **q**. This allowed him to rewrite perturbation theory precisely in the form of diagrams. Richard Feynman later gave an independent systematic derivation of these diagrams from a particle formalism, and they are now called Feynman diagrams. Each line of a diagram represents a particle propagating either backward or forward in time. This technique is the most widespread method of computing amplitudes in quantum field theory today.

Since this picture was first developed by Ernst Stueckelberg, and acquired its modern form in Feynman's work, it is called the *Feynman-Stueckelberg interpretation* of antiparticles to honor both scientists.

As a consequence of this interpretation, Villata argued that the assumption of antimatter as CPT-transformed matter would imply that the gravitational interaction between matter and antimatter is repulsive.[5]

2.5   See also

- Gravitational interaction of antimatter

- Parity, charge conjugation and time reversal symmetry.

- CP violations and the baryon asymmetry of the universe.

- Quantum field theory and the list of particles

- Baryogenesis

2.6   References

[1] http://news.nationalgeographic.com/news/2010/11/101118-antimatter-trapped-engines-bombs-nature-science-cern/

[2] Weinberg, Steve. *The quantum theory of fields, Volume 1 : Foundations.* p. 14. ISBN 0-521-55001-7.

[3] Dirac, Paul (1930). "A Theory of Electrons and Protons".*Proceedings of the Royal Society A***126**(801): 360–365.Bibcode:1 doi:10.1098/rspa.1930.0013.

[4] Sodickson, L.; W. Bowman; J. Stephenson (1961). "Single-Quantum Annihilation of Positrons". *Physical Review* **124** (6): 1851–1861. Bibcode:1961PhRv..124.1851S. doi:10.1103/PhysRev.124.1851.

[5] M. Villata, CPT symmetry and antimatter gravity in general relativity, 2011, EPL (Europhysics Letters) 94, 20001

- Feynman, R. P. (1987). "The reason for antiparticles". In R. P. Feynman and S. Weinberg. *The 1986 Dirac memorial lectures.* Cambridge University Press. ISBN 0-521-34000-4.

- Weinberg, S. (1995). *The Quantum Theory of Fields, Volume 1: Foundations.* Cambridge University Press. ISBN 0-521-55001-7.

Chapter 3

Positron emission

Positron emission or **beta plus decay** (β^+ decay) is a particular type of radioactive decay and a subtype of beta decay, in which a proton inside a radionuclide nucleus is converted into a neutron while releasing a positron and an electron neutrino (ν_e).[1] Positron emission is mediated by the weak force. The positron is a type of beta particle (β^+), the other beta particle being the electron (β^-) emitted from the β^- decay of a nucleus.

An example of positron emission (β^+ decay) is shown with magnesium-23 decaying into sodium-23:

> 23
> 12Mg → 23
> 11Na + e+ + ν
> e

Because positron emission decreases proton number relative to neutron number, positron decay happens typically in large "proton-rich" radionuclides. Positron decay results in nuclear transmutation, changing an atom of a chemical element into an atom of an element with an atomic number that is less by one unit.

Positron emission should not be confused with electron emission or beta minus decay (β^- decay), which occurs when a neutron turns into a proton and the nucleus emits an electron and an antineutrino.

Electron capture (sometimes called inverse beta decay) is also occasionally classified as a type of beta decay. In some ways, electron capture can be regarded as an equivalent to positron emission, since capture of an electron results in the same transmutation as emission of a positron. Electron capture occurs when electrons are available and requires less energy difference between parent and daughter, so occurs much more often in smaller atoms than positron emission does. Electron capture always competes with positron emission where the latter is seen, and in addition, occurs as the only type of beta decay in proton-rich nuclei when there is not enough decay energy to support positron emission.

3.1   Discovery of positron emission

In 1934 Frédéric and Irène Joliot-Curie bombarded aluminium with alpha particles to effect the nuclear reaction 4
2He + 27
13Al → 30
15P + 1
0n, and observed that the product isotope 30
15P emits a positron identical to those found in cosmic rays by Carl David Anderson in 1932.[2] This was the first example of β+ decay (positron emission). The Curies termed the phenomenon "artificial radioactivity," since 30
15P is a short-lived nuclide which does not exist in nature. The discovery of artificial radioactivity would be cited when the husband and wife team won the Nobel Prize.

3.2   Positron-emitting isotopes

Isotopes which undergo this decay and thereby emit positrons include carbon-11, potassium-40, nitrogen-13, oxygen-15, aluminium-26, sodium-22, fluorine-18, and iodine-121. As an example, the following equation describes the beta plus decay of carbon-11 to boron−11, emitting a positron and a neutrino:

3.3   Emission mechanism

Inside protons and neutrons, there are fundamental particles called quarks. The two most common types of quarks are up quarks, which have a charge of $+^2/_3$ and down quarks, with a $-^1/_3$ charge. Quarks arrange themselves in sets of three such that they make protons and neutrons. In a proton, whose charge is +1, there are two up quarks and one down quark. Neutrons, with no charge, have one up quark and two down quarks. Via the weak interaction, quarks can change flavor from down to up, resulting in electron emission. Positron emission happens when an up quark changes into a down quark.[3]

Nuclei which decay by positron emission may also decay by electron capture. For low-energy decays, electron capture is energetically favored by $2m_e c^2 = 1.022$ MeV, since the final state has an electron removed rather than a positron added. As the energy of the decay goes up, so does the branching ratio towards positron emission. However, if the energy difference is less than $2m_e c^2$, then positron emission cannot occur and electron capture is the sole decay mode. Certain isotopes (for instance, 7Be) are stable in galactic cosmic rays, because the electrons are stripped away and the decay energy is too small for positron emission.

3.4   Application

These isotopes are used in positron emission tomography, a technique used for medical imaging. Note that the energy emitted depends on the isotope that is decaying; the figure of 0.96 MeV applies only to the decay of carbon-11. Isotopes which increase in mass under the conversion of a proton to a neutron, or which decrease in mass by less than $2m_e$, cannot spontaneously decay by positron emission.

The short-lived positron emitting isotopes ^{11}C, ^{13}N, ^{15}O and ^{18}F used for positron emission tomography are typically produced by proton irradiation of natural or enriched targets.[4][5]

3.5   References

[1] The University of North Carolina at Chapel Hill. "Nuclear Chemistry". Retrieved 2012-06-14.

[2] I. Curie and F. Joliot, *C. R. Acad. Sci.* 198, 254 (1934)

[3] How it works:Positron emission

[4] Positron Emission Tomography Imaging at the University of British Columbia (accessed 11 May 2012)

[5] Ledingham, K W D; McKenna, P; McCanny, T; Shimizu, S; Yang, J M; Robson, L; Zweit, J; Gillies, J M; Bailey, J; Chimon, G N; Clarke, R J; Neely, D; Norreys, P A; Collier, J L; Singhal, R P; Wei, M S; Mangles, S P D; Nilson, P; Krushelnick, K; Zepf, M (2004). "High power laser production of short-lived isotopes for positron emission tomography". *Journal of Physics D: Applied Physics* 37 (16): 2341. Bibcode:2004JPhD...37.2341L. doi:10.1088/0022-3727/37/16/019.

3.6 External links

-  **The LIVEChart of Nuclides - IAEA** with filter on β+ decay

Chapter 4

Positron emission tomography

Positron emission tomography (**PET**)[1] is a nuclear medicine, functional imaging technique that produces a three-dimensional image of functional processes in the body. The system detects pairs of gamma rays emitted indirectly by a positron-emitting radionuclide (tracer), which is introduced into the body on a biologically active molecule. Three-dimensional images of tracer concentration within the body are then constructed by computer analysis. In modern PET-CT scanners, three dimensional imaging is often accomplished with the aid of a CT X-ray scan performed on the patient during the same session, in the same machine.

If the biologically active molecule chosen for PET is fluorodeoxyglucose (FDG), an analogue of glucose, the concentrations of tracer imaged will indicate tissue metabolic activity as it corresponds to the regional glucose uptake. Use of this tracer to explore the possibility of cancer metastasis (i.e., spreading to other sites) is the most common type of PET scan in standard medical care (90% of current scans). However, on a minority basis, many other radioactive tracers are used in PET to image the tissue concentration of other types of molecules of interest. One of the disadvantages of PET scanners is their operating cost.[2]

4.1 Uses

PET is both a medical and research tool. It is used heavily in clinical oncology (medical imaging of tumors and the search for metastases), and for clinical diagnosis of certain diffuse brain diseases such as those causing various types of dementias. PET is also an important research tool to map normal human brain and heart function, and support drug development.

PET is also used in pre-clinical studies using animals, where it allows repeated investigations into the same subjects. This is particularly valuable in cancer research, as it results in an increase in the statistical quality of the data (subjects can act as their own control) and substantially reduces the numbers of animals required for a given study.

Alternative methods of scanning include x-ray computed tomography (CT), magnetic resonance imaging (MRI) and functional magnetic resonance imaging (fMRI), ultrasound and single-photon emission computed tomography (SPECT).

While some imaging scans such as CT and MRI isolate organic anatomic changes in the body, PET and SPECT are capable of detecting areas of molecular biology detail (even prior to anatomic change). PET scanning does this using radiolabelled molecular probes that have different rates of uptake depending on the type and function of tissue involved. Changing of regional blood flow in various anatomic structures (as a measure of the injected positron emitter) can be visualized and relatively quantified with a PET scan.

PET imaging is best performed using a dedicated PET scanner. However, it is possible to acquire PET images using a conventional dual-head gamma camera fitted with a coincidence detector. The quality of gamma-camera PET is considerably lower, and acquisition is slower. However, for institutions with low demand for PET, this may allow on-site imaging, instead of referring patients to another center, or relying on a visit by a mobile scanner.

PET is a valuable technique for some diseases and disorders, because it is possible to target the radio-chemicals used for

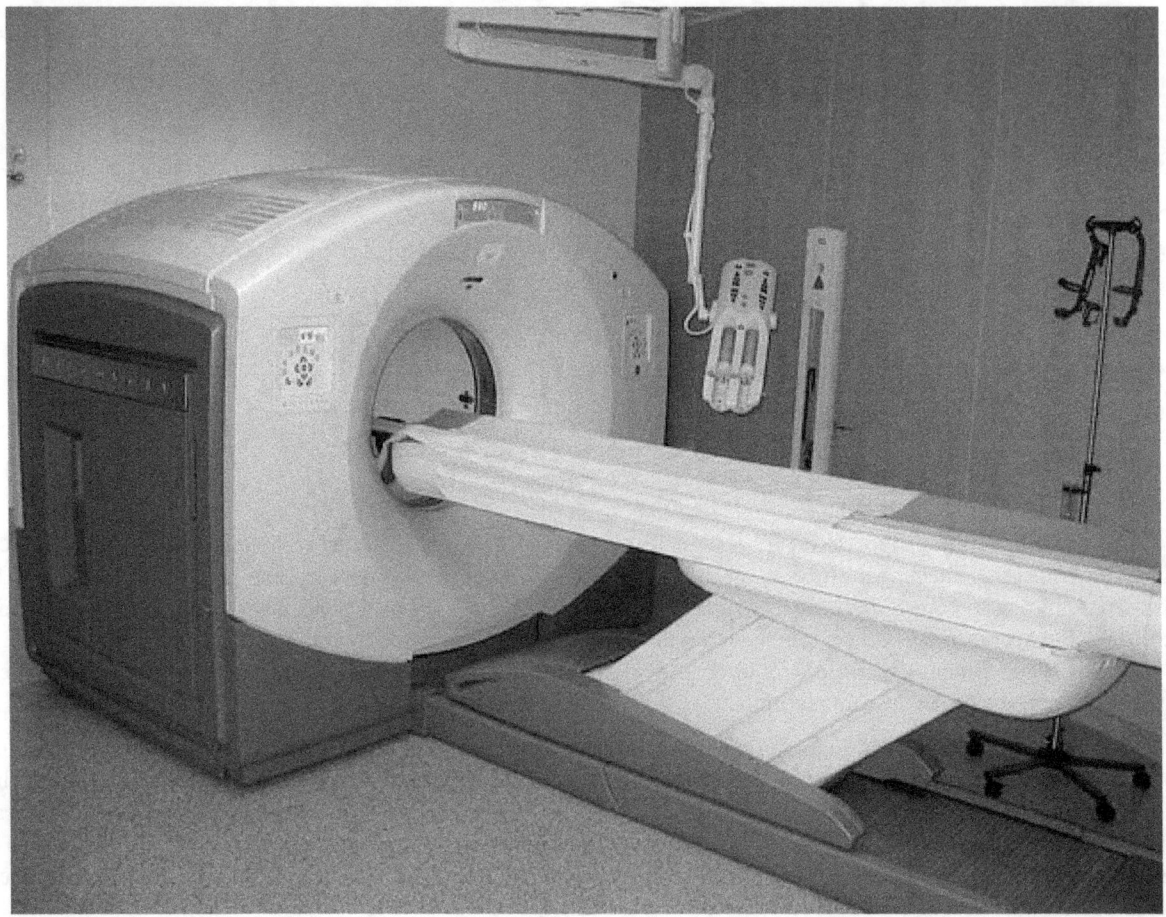

PET/CT-System with 16-slice CT; the ceiling mounted device is an injection pump for CT contrast agent

particular bodily functions.

4.1.1 Oncology

Oncology: PET scanning with the tracer fluorine-18 (F-18) fluorodeoxyglucose (FDG), called FDG-PET, is widely used in clinical oncology. This tracer is a glucose analog that is taken up by glucose-using cells and phosphorylated by hexokinase (whose mitochondrial form is greatly elevated in rapidly growing malignant tumours). A typical dose of FDG used in an oncological scan has an effective radiation dose of 14 mSv.[3] Because the oxygen atom that is replaced by F-18 to generate FDG is required for the next step in glucose metabolism in all cells, no further reactions occur in FDG. Furthermore, most tissues (with the notable exception of liver and kidneys) cannot remove the phosphate added by hexokinase. This means that FDG is trapped in any cell that takes it up, until it decays, since phosphorylated sugars, due to their ionic charge, cannot exit from the cell. This results in intense radiolabeling of tissues with high glucose uptake, such as the brain, the liver, and most cancers. As a result, FDG-PET can be used for diagnosis, staging, and monitoring treatment of cancers, particularly in Hodgkin's lymphoma, non-Hodgkin lymphoma, and lung cancer. Many other types of solid tumors will be found to be very highly labeled on a case-by-case basis—a fact that becomes especially useful in searching for tumor metastasis, or for recurrence after a known highly active primary tumor is removed. Because individual PET scans are more expensive than "conventional" imaging with computed tomography (CT) and magnetic resonance imaging (MRI), expansion of FDG-PET in cost-constrained health services will depend on proper health technology assessment; this problem is a difficult one because structural and functional imaging often cannot be directly compared, as they provide different information. Oncology scans using FDG make up over 90% of all PET scans in current practice .

A few other isotopes and radiotracers are slowly being introduced into oncology for specific purposes. For example,

11C-Metomidate has been used to detect tumors of adrenocortical origin.[4][5] Also, FDOPA PET-CT, in centers which offer it, has proven to be a more sensitive alternative to finding, and also localizing pheochromocytoma than the MIBG scan.[6][7][8]

4.1.2   Neuroimaging

Main article: Brain positron emission tomography

Neurology: PET neuroimaging is based on an assumption that areas of high radioactivity are associated with brain activity. What is actually measured indirectly is the flow of blood to different parts of the brain, which is, in general, believed to be correlated, and has been measured using the tracer oxygen−15. However, because of its 2-minute half-life, O-15 must be piped directly from a medical cyclotron for such uses, which is difficult. In practice, since the brain is normally a rapid user of glucose, and since brain pathologies such as Alzheimer's disease greatly decrease brain metabolism of both glucose and oxygen in tandem, standard FDG-PET of the brain, which measures regional glucose use, may also be successfully used to differentiate Alzheimer's disease from other dementing processes, and also to make early diagnosis of Alzheimer's disease. The advantage of FDG-PET for these uses is its much wider availability. PET imaging with FDG can also be used for localization of seizure focus: A seizure focus will appear as hypometabolic during an interictal scan. Several radiotracers (i.e. radioligands) have been developed for PET that are ligands for specific neuroreceptor subtypes such as [11C] raclopride, [18F] fallypride and [18F] desmethoxyfallypride for dopamine D2/D3 receptors, [11C] McN 5652 and [11C] DASB for serotonin transporters, [18F] Mefway for serotonin 5HT1A receptors, [18F] Nifene for nicotinic acetylcholine receptors or enzyme substrates (e.g. 6-FDOPA for the AADC enzyme). These agents permit the visualization of neuroreceptor pools in the context of a plurality of neuropsychiatric and neurologic illnesses. The development of a number of novel probes for noninvasive, in vivo PET imaging of neuroaggregate in human brain has brought amyloid imaging to the doorstep of clinical use. The earliest amyloid imaging probes included 2-(1-{6-[(2-[18F]fluoroethyl)(methyl)amino]−2-naphthyl}ethylidene)malononitrile ([18F]FDDNP)[9] developed at the University of California, Los Angeles and N-methyl-[11C]2-(4'-methylaminophenyl)−6-hydroxybenzothiazole[10] (termed Pittsburgh compound B) developed at the University of Pittsburgh. These amyloid imaging probes permit the visualization of amyloid plaques in the brains of Alzheimer's patients and could assist clinicians in making a positive clinical diagnosis of AD pre-mortem and aid in the development of novel anti-amyloid therapies. [11C]PMP (N-[11C]methylpiperidin-4-yl propionate) is a novel radiopharmaceutical used in PET imaging to determine the activity of the acetylcholinergic neurotransmitter system by acting as a substrate for acetylcholinesterase. Post-mortem examination of AD patients have shown decreased levels of acetylcholinesterase. [11C]PMP is used to map the acetylcholinesterase activity in the brain, which could allow for pre-mortem diagnosis of AD and help to monitor AD treatments.[11] Avid Radiopharmaceuticals of Philadelphia has developed a compound called 18F-AV-45 that uses the longer-lasting radionuclide fluorine-18 to detect amyloid plaques using PET scans.[12]

2. Neuropsychology / Cognitive neuroscience: To examine links between specific psychological processes or disorders and brain activity.

3. Psychiatry: Numerous compounds that bind selectively to neuroreceptors of interest in biological psychiatry have been radiolabeled with C-11 or F-18. Radioligands that bind to dopamine receptors (D1,[13] D2 receptor,[14][15] reuptake transporter), serotonin receptors (5HT1A, 5HT2A, reuptake transporter) opioid receptors (mu) and other sites have been used successfully in studies with human subjects. Studies have been performed examining the state of these receptors in patients compared to healthy controls in schizophrenia, substance abuse, mood disorders and other psychiatric conditions.

4. Stereotactic surgery and radiosurgery: PET-image guided surgery facilitates treatment of intracranial tumors, arteriovenous malformations and other surgically treatable conditions.[16]

4.1.3 Cardiology

Main article: Cardiac PET

Cardiology, atherosclerosis and vascular disease study: In clinical cardiology, FDG-PET can identify so-called "hibernating myocardium", but its cost-effectiveness in this role versus SPECT is unclear. FDG-PET imaging of atherosclerosis to detect patients at risk of stroke is also feasible and can help test the efficacy of novel anti-atherosclerosis therapies.[17]

4.1.4 Infectious Diseases

Imaging infections with molecular imaging technologies can improve diagnosis and treatment follow-up. PET has been widely used to image bacterial infections clinically by using fluorodeoxyglucose (FDG) to identify the infection-associated inflammatory response.

Recently, three different PET contrast agents have been developed to image bacterial infections in vivo: [^{18}F]maltose,[18] [^{18}F]maltohexaose and [^{18}F]2-fluorodeoxysorbitol (FDS).[19] FDS has also the added benefit of being able to target only Enterobacteriaceae.

4.1.5 Pharmacokinetics

Pharmacokinetics: In pre-clinical trials, it is possible to radiolabel a new drug and inject it into animals. Such scans are referred to as biodistribution studies. The uptake of the drug, the tissues in which it concentrates, and its eventual elimination, can be monitored far more quickly and cost effectively than the older technique of killing and dissecting the animals to discover the same information. Much more commonly, however, drug occupancy at a purported site of action can be inferred indirectly by competition studies between unlabeled drug and radiolabeled compounds known apriori to bind with specificity to the site. A single radioligand can be used this way to test many potential drug candidates for the same target. A related technique involves scanning with radioligands that compete with an endogenous (naturally occurring) substance at a given receptor to demonstrate that a drug causes the release of the natural substance.

The following is an excerpt from an article by Harvard University staff writer Peter Reuell, featured in HarvardScience, part of the online version of the Harvard Gazette newspaper, which discusses research by the team of Harvard Associate Professor of Organic Chemistry and Chemical Biology Tobias Ritter: "A new chemical process ... may increase the utility of positron emission tomography (PET) in creating real-time 3-D images of chemical activity occurring inside the body. This new work ... holds out the tantalizing possibility of using PET scans to peer into a number of functions inside animals and humans by simplifying the process of using "tracer" molecules to create the 3-D images." (by creating a novel electrophilic fluorination reagent as an intermediate molecule; the research could be used in drug development).[20]

4.1.6 Small animal imaging

PET technology for small animal imaging: A miniature PE tomograph has been constructed that is small enough for a fully conscious and mobile rat to wear on its head while walking around.[21] This RatCAP (Rat Conscious Animal PET) allows animals to be scanned without the confounding effects of anesthesia. PET scanners designed specifically for imaging rodents, often referred to as microPET, as well as scanners for small primates are marketed for academic and pharmaceutical research.

4.1.7 Musculo-skeletal imaging

Musculoskeletal imaging: PET has been shown to be a feasible technique for studying skeletal muscles during exercises like walking.[22] One of the main advantages of using PET is that it can also provide muscle activation data about deeper lying muscles such as the vastus intermedialis and the gluteus minimus, as compared to other muscle studying techniques like electromyography, which can be used only on superficial muscles (i.e., directly under the skin). A clear disadvantage, however, is that PET provides no timing information about muscle activation, because it has to be measured after the exercise is completed. This is due to the time it takes for FDG to accumulate in the activated muscles.

4.2   Safety

PET scanning is non-invasive, but it does involve exposure to ionizing radiation.[2]

18F-FDG, which is now the standard radiotracer used for PET neuroimaging and cancer patient management,[23] has an effective radiation dose of 14 mSv.[3]

The amount of radiation in 18F-FDG is similar to the effective dose of spending one year in Denver, CO (12.4 mSv/year).[24] For comparison, radiation dosage for other medical procedures range from 0.02 mSv for a chest x-ray and 6.5–8 mSv for a CT scan of the chest.[25][26] Average civil aircrews are exposed to 3 mSv/year,[27] and the whole body occupational dose limit for Nuclear Energy Workers in the USA is 50mSv/year.[28] For scale, see Orders of magnitude (radiation).

For PET-CT scanning, the radiation exposure may be substantial—around 23–26 mSv (for a 70 kg person—dose is likely to be higher for higher body weights).[29]

4.3   Descriptions

4.3.1   Operation

To conduct the scan, a short-lived radioactive tracer isotope is injected into the living subject (usually into blood circulation). Each tracer atom has been chemically incorporated into a biologically active molecule. There is a waiting period while the active molecule becomes concentrated in tissues of interest; then the subject is placed in the imaging scanner. The molecule most commonly used for this purpose is F-18 labeled fluorodeoxyglucose (FDG), a sugar, for which the waiting period is typically an hour. During the scan, a record of tissue concentration is made as the tracer decays.

As the radioisotope undergoes positron emission decay (also known as positive beta decay), it emits a positron, an antiparticle of the electron with opposite charge. The emitted positron travels in tissue for a short distance (typically less than 1 mm, but dependent on the isotope[30]), during which time it loses kinetic energy, until it decelerates to a point where it can interact with an electron.[31] The encounter annihilates both electron and positron, producing a pair of annihilation (gamma) photons moving in approximately opposite directions. These are detected when they reach a scintillator in the scanning device, creating a burst of light which is detected by photomultiplier tubes or silicon avalanche photodiodes (Si APD). The technique depends on simultaneous or coincident detection of the pair of photons moving in approximately opposite directions (they would be exactly opposite in their center of mass frame, but the scanner has no way to know this, and so has a built-in slight direction-error tolerance). Photons that do not arrive in temporal "pairs" (i.e. within a timing-window of a few nanoseconds) are ignored.

4.3.2   Localization of the positron annihilation event

The most significant fraction of electron–positron annihilations results in two 511 keV gamma photons being emitted at almost 180 degrees to each other; hence, it is possible to localize their source along a straight line of coincidence (also called the **line of response**, or **LOR**). In practice, the LOR has a non-zero width as the emitted photons are not exactly 180 degrees apart. If the resolving time of the detectors is less than 500 picoseconds rather than about 10 nanoseconds, it is possible to localize the event to a segment of a chord, whose length is determined by the detector timing resolution. As the timing resolution improves, the signal-to-noise ratio (SNR) of the image will improve, requiring fewer events to achieve the same image quality. This technology is not yet common, but it is available on some new systems.[32]

4.3.3 Image reconstruction using coincidence statistics

A technique much like the reconstruction of computed tomography (CT) and single-photon emission computed tomography (SPECT) data is more commonly used, although the data set collected in PET is much poorer than CT, so reconstruction techniques are more difficult (see Image reconstruction of PET).

Using statistics collected from tens of thousands of coincidence events, a set of simultaneous equations for the total activity of each parcel of tissue along many LORs can be solved by a number of techniques, and, thus, a map of radioactivities as a function of location for parcels or bits of tissue (also called voxels) can be constructed and plotted. The resulting map shows the tissues in which the molecular tracer has become concentrated, and can be interpreted by a nuclear medicine physician or radiologist in the context of the patient's diagnosis and treatment plan.

4.3.4 Combination of PET with CT or MRI

Main articles: PET-CT and PET-MRI

PET scans are increasingly read alongside CT or magnetic resonance imaging (MRI) scans, with the combination (called "co-registration") giving both anatomic and metabolic information (i.e., what the structure is, and what it is doing biochemically). Because PET imaging is most useful in combination with anatomical imaging, such as CT, modern PET scanners are now available with integrated high-end multi-detector-row CT scanners (so-called "PET-CT"). Because the two scans can be performed in immediate sequence during the same session, with the patient not changing position between the two types of scans, the two sets of images are more-precisely registered, so that areas of abnormality on the PET imaging can be more perfectly correlated with anatomy on the CT images. This is very useful in showing detailed views of moving organs or structures with higher anatomical variation, which is more common outside the brain.

At the Jülich Institute of Neurosciences and Biophysics, the world's largest PET-MRI device began operation in April 2009: a 9.4-tesla magnetic resonance tomograph (MRT) combined with a positron emission tomograph (PET). Presently, only the head and brain can be imaged at these high magnetic field strengths.[33]

For brain imaging, registration of CT, MRI and PET scans may be accomplished without the need for an integrated PET-CT or PET-MRI scanner by using a device known as the N-localizer.[34][35][36][37][38][39][40][16]

4.3.5 Radionuclides and radiotracers

Main articles: List of PET radiotracers and Fludeoxyglucose

Radionuclides used in PET scanning are typically isotopes with short half-lives [2] such as carbon-11 (~20 min), nitrogen-13 (~10 min), oxygen-15 (~2 min), fluorine-18 (~110 min), gallium-68 (~67 min), zirconium-89 (~78.41 hours), or rubidium-82(~1.27 min). These radionuclides are incorporated either into compounds normally used by the body such as glucose (or glucose analogues), water, or ammonia, or into molecules that bind to receptors or other sites of drug action. Such labelled compounds are known as radiotracers. PET technology can be used to trace the biologic pathway of any compound in living humans (and many other species as well), provided it can be radiolabeled with a PET isotope. Thus, the specific processes that can be probed with PET are virtually limitless, and radiotracers for new target molecules and processes are continuing to be synthesized; as of this writing there are already dozens in clinical use and hundreds applied in research. At present, however, by far the most commonly used radiotracer in clinical PET scanning is fluorodeoxyglucose (also called FDG or fludeoxyglucose), an analogue of glucose that is labeled with fluorine-18. This radiotracer is used in essentially all scans for oncology and most scans in neurology, and thus makes up the large majority of all of the radiotracer (> 95%) used in PET and PET-CT scanning.

Due to the short half-lives of most positron-emitting radioisotopes, the radiotracers have traditionally been produced using a cyclotron in close proximity to the PET imaging facility. The half-life of fluorine-18 is long enough that radiotracers labeled with fluorine-18 can be manufactured commercially at offsite locations and shipped to imaging centers. Recently rubidium−82 generators have become commercially available.[41] These contain strontium-82, which decays by electron capture to produce positron-emitting rubidium-82.

4.3.6 Limitations

The minimization of radiation dose to the subject is an attractive feature of the use of short-lived radionuclides. Besides its established role as a diagnostic technique, PET has an expanding role as a method to assess the response to therapy, in particular, cancer therapy,[42] where the risk to the patient from lack of knowledge about disease progress is much greater than the risk from the test radiation.

Limitations to the widespread use of PET arise from the high costs of cyclotrons needed to produce the short-lived radionuclides for PET scanning and the need for specially adapted on-site chemical synthesis apparatus to produce the radiopharmaceuticals after radioisotope preparation. Organic radiotracer molecules that will contain a positron-emitting radioisotope cannot be synthesized first and then the radioisotope prepared within them, because bombardment with a cyclotron to prepare the radioisotope destroys any organic carrier for it. Instead, the isotope must be prepared first, then afterward, the chemistry to prepare any organic radiotracer (such as FDG) accomplished very quickly, in the short time before the isotope decays. Few hospitals and universities are capable of maintaining such systems, and most clinical PET is supported by third-party suppliers of radiotracers that can supply many sites simultaneously. This limitation restricts clinical PET primarily to the use of tracers labelled with fluorine-18, which has a half-life of 110 minutes and can be transported a reasonable distance before use, or to rubidium-82 (used as rubidium-82 chloride) with a half-life of 1.27 minutes, which is created in a portable generator and is used for myocardial perfusion studies. Nevertheless, in recent years a few on-site cyclotrons with integrated shielding and "hot labs" (automated chemistry labs that are able to work with radioisotopes) have begun to accompany PET units to remote hospitals. The presence of the small on-site cyclotron promises to expand in the future as the cyclotrons shrink in response to the high cost of isotope transportation to remote PET machines.[43] In recent years the shortage of PET scans has been alleviated in the US, as rollout of radiopharmacies to supply radioisotopes has grown 30%/year.[44]

Because the half-life of fluorine-18 is about two hours, the prepared dose of a radiopharmaceutical bearing this radionuclide will undergo multiple half-lives of decay during the working day. This necessitates frequent recalibration of the remaining dose (determination of activity per unit volume) and careful planning with respect to patient scheduling.

4.3.7 Image reconstruction

The raw data collected by a PET scanner are a list of 'coincidence events' representing near-simultaneous detection (typically, within a window of 6 to 12 nanoseconds of each other) of annihilation photons by a pair of detectors. Each coincidence event represents a line in space connecting the two detectors along which the positron emission occurred (i.e., the line of response (LOR)). Modern systems with a higher time resolution (roughly 3 nanoseconds) also use a technique (called "Time-of-flight") where they more precisely decide the difference in time between the detection of the two photons and can thus localize the point of origin of the annihilation event between the two detectors to within 10 cm.

Coincidence events can be grouped into projection images, called sinograms. The sinograms are sorted by the angle of each view and tilt (for 3D images). The sinogram images are analogous to the projections captured by computed tomography (CT) scanners, and can be reconstructed in a similar way. However, the statistics of the data are much worse than those obtained through transmission tomography. A normal PET data set has millions of counts for the whole acquisition, while the CT can reach a few billion counts. This contributes to PET images appearing "noisier" than CT. Two major sources of noise in PET are scatter (a detected pair of photons, at least one of which was deflected from its original path by interaction with matter in the field of view, leading to the pair being assigned to an incorrect LOR) and random events (photons originating from two different annihilation events but incorrectly recorded as a coincidence pair because their arrival at their respective detectors occurred within a coincidence timing window).

In practice, considerable pre-processing of the data is required—correction for random coincidences, estimation and subtraction of scattered photons, detector dead-time correction (after the detection of a photon, the detector must "cool down" again) and detector-sensitivity correction (for both inherent detector sensitivity and changes in sensitivity due to angle of incidence).

Filtered back projection (FBP) has been frequently used to reconstruct images from the projections. This algorithm has the advantage of being simple while having a low requirement for computing resources. However, shot noise in the raw data is prominent in the reconstructed images and areas of high tracer uptake tend to form streaks across the image. Also, FBP treats the data deterministically—it does not account for the inherent randomness associated with PET data, thus requiring all the pre-reconstruction corrections described above.

Iterative expectation-maximization algorithms are now the preferred method of reconstruction. These algorithms compute an estimate of the likely distribution of annihilation events that led to the measured data, based on statistical principles. The advantage is a better noise profile and resistance to the streak artifacts common with FBP, but the disadvantage is higher computer resource requirements.[45]

Recent research has shown that Bayesian methods that involve a Poisson likelihood function and an appropriate prior probability (e.g., a smoothing prior leading to total variation regularization or a Laplacian distribution leading to ℓ_1 - based regularization in a wavelet or other domain) may yield superior performance to expectation-maximization-based methods which involve a Poisson likelihood function but do not involve such a prior.[46][47][48]

Attenuation correction: Attenuation occurs when photons emitted by the radiotracer inside the body are absorbed by intervening tissue between the detector and the emission of the photon. As different LORs must traverse different thicknesses of tissue, the photons are attenuated differentially. The result is that structures deep in the body are reconstructed as having falsely low tracer uptake. Contemporary scanners can estimate attenuation using integrated x-ray CT equipment, however earlier equipment offered a crude form of CT using a gamma ray (positron emitting) source and the PET detectors.

While attenuation-corrected images are generally more faithful representations, the correction process is itself susceptible to significant artifacts. As a result, both corrected and uncorrected images are always reconstructed and read together.

2D/3D reconstruction: Early PET scanners had only a single ring of detectors, hence the acquisition of data and subsequent reconstruction was restricted to a single transverse plane. More modern scanners now include multiple rings, essentially forming a cylinder of detectors.

There are two approaches to reconstructing data from such a scanner: 1) treat each ring as a separate entity, so that only coincidences within a ring are detected, the image from each ring can then be reconstructed individually (2D reconstruction), or 2) allow coincidences to be detected between rings as well as within rings, then reconstruct the entire volume together (3D).

3D techniques have better sensitivity (because more coincidences are detected and used) and therefore less noise, but are more sensitive to the effects of scatter and random coincidences, as well as requiring correspondingly greater computer resources. The advent of sub-nanosecond timing resolution detectors affords better random coincidence rejection, thus favoring 3D image reconstruction.

4.4 History

The concept of emission and transmission tomography was introduced by David E. Kuhl, Luke Chapman and Roy Edwards in the late 1950s. Their work later led to the design and construction of several tomographic instruments at the University of Pennsylvania. Tomographic imaging techniques were further developed by Michel Ter-Pogossian, Michael E. Phelps, Edward J. Hoffman and others at Washington University School of Medicine.[49][50]

Work by Gordon Brownell, Charles Burnham and their associates at the Massachusetts General Hospital beginning in the 1950s contributed significantly to the development of PET technology and included the first demonstration of annihilation radiation for medical imaging.[51] Their innovations, including the use of light pipes and volumetric analysis, have been important in the deployment of PET imaging. In 1961, James Robertson and his associates at Brookhaven National Laboratory built the first single-plane PET scan, nicknamed the "head-shrinker."[52]

One of the factors most responsible for the acceptance of positron imaging was the development of radiopharmaceuticals. In particular, the development of labeled 2-fluorodeoxy-D-glucose (2FDG) by the Brookhaven group under the direction of Al Wolf and Joanna Fowler was a major factor in expanding the scope of PET imaging.[53] The compound was first administered to two normal human volunteers by Abass Alavi in August 1976 at the University of Pennsylvania. Brain images obtained with an ordinary (non-PET) nuclear scanner demonstrated the concentration of FDG in that organ. Later, the substance was used in dedicated positron tomographic scanners, to yield the modern procedure.

The logical extension of positron instrumentation was a design using two 2-dimensional arrays. PC-I was the first instrument using this concept and was designed in 1968, completed in 1969 and reported in 1972. The first applications of

PC-I in tomographic mode as distinguished from the computed tomographic mode were reported in 1970.[54] It soon became clear to many of those involved in PET development that a circular or cylindrical array of detectors was the logical next step in PET instrumentation. Although many investigators took this approach, James Robertson[55] and Zang-Hee Cho[56] were the first to propose a ring system that has become the prototype of the current shape of PET.

The PET-CT scanner, attributed to Dr. David Townsend and Dr. Ronald Nutt, was named by TIME Magazine as the medical invention of the year in 2000.[57]

4.5   Cost

As of August 2008, Cancer Care Ontario reports that the current average incremental cost to perform a PET scan in the province is $1,000–$1,200 per scan. This includes the cost of the radiopharmaceutical and a stipend for the physician reading the scan.[58]

4.6   Research

There are ongoing studies into using potassium 40 as this decays with a positron channel if a way could be found to enrich the isotope into a pure form in order to scan human neural pathways to study neurodegenerative diseases such as Alzheimers disease.

4.7   See also

- Diffuse optical imaging

- Hot cell (Equipment used to produce the radiopharmaceuticals used in PET)

- Molecular Imaging

4.8   References

[1] Bailey, D.L; D.W. Townsend; P.E. Valk; M.N. Maisey (2005). *Positron Emission Tomography: Basic Sciences*. Secaucus, NJ: Springer-Verlag. ISBN 1-85233-798-2.

[2] Carlson, Neil (January 22, 2012). *Physiology of Behavior*. Methods and Strategies of Research. 11th edition. Pearson. p. 151. ISBN 0205239390.

[3] Exposure fact sheet Health Physics Society

[4] Khan TS; Sundin A; Juhlin C; Långström B; et al. (2003). "11C-metomidate PET imaging of adrenocortical cancer". *European Journal of Nuclear Medicine and Molecular Imaging* **30** (3): 403–410. doi:10.1007/s00259-002-1025-9. PMID 12634969.

[5] Minn H; Salonen A; Friberg J; Roivainen A; et al. (June 2004). "Imaging of adrenal incidentalomas with PET using (11)C-metomidate and (18)F-FDG". *J. Nucl. Med.* **45** (6): 972–9. PMID 15181132.

[6] full text of early article on FDOPA PET for pheochromocytoma

[7] imaging overview

[8] Luster M; Karges W; Zeich K; Pauls S; et al. (2010). "Clinical value of 18F-fluorodihydroxyphenylalanine positron emission tomography/computed tomography (18F-DOPA PET/CT) for detecting pheochromocytoma". *Eur. J. Nucl. Med. Mol. Imaging* **37** (3): 484–93. doi:10.1007/s00259-009-1294-7. PMID 19862519.

[9] Agdeppa ED; Kepe V; Liu J; Flores-Torres S; et al. (2001). "Binding characteristics of radiofluorinated 6-dialkylamino-2-naphthylethylidene derivatives as positron emission tomography imaging probes for beta-amyloid plaques in Alzheimer's disease" (PDF). *J Neurosci* **21** (24): RC189(1–5). PMID 11734604.

[10] Mathis CA; Bacskai BJ; Kajdasz ST; McLellan ME; et al. (2002). "A lipophilic thioflavin-T derivative for positron emission tomography (PET) imaging of amyloid in brain". *Bioorg. Med. Chem. Lett.* **12** (3): 295–8. doi:10.1016/S0960-894X(01)00734-X. PMID 11814781.

[11] Kuhl DE; Koeppe RA; Minoshima S; Snyder SE; et al. (March 1999). "In vivo mapping of cerebral acetylcholinesterase activity in aging and Alzheimer's disease". *Neurology* **52** (4): 691–9. doi:10.1212/wnl.52.4.691. PMID 10078712.

[12] Kolata, Gina. "Promise Seen for Detection of Alzheimer's", *The New York Times*, June 23, 2010. Accessed June 23, 2010.

[13] Catafau AM; Searle GE; Bullich S; Gunn RN; et al. (2010). "Imaging cortical dopamine D1 receptors using 11C NNC112 and ketanserin blockade of the 5-HT 2A receptors". *J Cereb Blood Flow Metab* **30** (5): 985–93. doi:10.1038/jcbfm.2009.269. PMID 20029452.

[14] Mukherjee J; Christian BT; Dunigan KA; Shi B; et al. (2002). "Brain imaging of 18F-fallypride in normal volunteers: Blood analysis, distribution, test-retest studies, and preliminary assessment of sensitivity to aging effects on dopamine D-2/D-3 receptors". *Synapse* **46** (3): 170–88. doi:10.1002/syn.10128. PMID 12325044.

[15] Buchsbaum MS; Christian BT; Lehrer DS; Narayanan TK; et al. (2006). "D2/D3 dopamine receptor binding with F-18fallypride in thalamus and cortex of patients with schizophrenia". *Schizophrenia research* **85** (1–3): 232–44. doi:10.1016/j.schres.2006.03.042.PMID16713185.

[16] Levivier M; Massager N; Wikler D; Lorenzoni J; et al. (July 2004). "Use of stereotactic PET images in dosimetry planning of radiosurgery for brain tumors: clinical experience and proposed classification". *Journal of Nuclear Medicine* **45** (7): 1146–1154. PMID 15235060.

[17] Rudd JH; Warburton EA; Fryer TD; Jones HA; et al. (2002). "Imaging atherosclerotic plaque inflammation with [18F]-fluorodeoxyglucose positron emission tomography". *Circulation* **105** (23): 2708–11. doi:10.1161/01.CIR.0000020548.60110.76. PMID 12057982.

[18] Gowrishankar, G.; Namavari, M.; Jouannot, E. B.; Hoehne, A.; et al. (2014). "Investigation of 6-[18F]-fluoromaltose as a novel PET tracer for imaging bacterial infection". *PLoS ONE* **9** (9): e107951. doi:10.1371/journal.pone.0107951. PMC 4171493. PMID 25243851.

[19] Weinstein EA; Ordonez AA; DeMarco VP; Murawski AM; et al. (2014). "Imaging Enterobacteriaceae infection in vivo with 18F-fluorodeoxysorbitol positron emission tomography". *Science Translational Medicine* **6** (259): 259ra146. doi:10.1126/scitranslmed.3009815.PMC4327834.PMID25338757.

[20] "Tracing biological pathways | Harvard Gazette". News.harvard.edu. Retrieved 2012-08-13.

[21] Rat Conscious Animal PET

[22] Oi N; Iwaya T; Itoh M; Yamaguchi K; et al. (2003). "FDG-PET imaging of lower extremity muscular activity during level walking". *J Orthop Sci* **8** (1): 55–61. doi:10.1007/s007760300009. PMID 12560887.

[23] Kelloff GJ; Hoffman JM; Johnson B; Scher HI; et al. (Apr 2005). "Progress and promise of FDG-PET imaging for cancer patient management and oncologic drug development". *Clin. Cancer Res.* **11** (8): 2785–808. doi:10.1158/1078-0432.CCR-04-2626. PMID 15837727.

[24] Background Radiation in Denver, Institute for Science and International Security

[25] Managing Patient Does, ICRP, 30 October 2009.

[26] de Jong PA; Tiddens HA; Lequin MH; Robinson TE; et al. (May 2008). "Estimation of the radiation dose from CT in cystic fibrosis". *Chest* **133** (5): 1289–91; author6 reply 1290–1. doi:10.1378/chest.07-2840. PMID 18460535.

[27] Chapter 9 Occupational Exposure to Radiation, IAEA

[28] Information for Radiation Workers , Nuclear Regulatory Commission

[29] Brix G; Lechel U; Glatting G; Ziegler SI; et al. (April 2005). "Radiation exposure of patients undergoing whole-body dual-modality 18F-FDG PET/CT examinations". *J. Nucl. Med.* **46** (4): 608–13. PMID 15809483.

[30] Michael E. Phelps (2006). *PET: physics, instrumentation, and scanners.* Springer. pp. 8–10. ISBN 0-387-34946-4.

[31] "PET Imaging". GE Healthcare. Archived from the original on 2012-02-05.

[32] "Invitation to Cover: Advancements in "Time-of-Flight" Technology Make New PET/CT Scanner at Penn a First in the World". University of Pennsylvania. June 15, 2006. Retrieved February 22, 2010.

[33] "A Close Look Into the Brain". Jülich Research Centre. 7 March 2014. Retrieved 2015-04-14.

[34] Brown RA; Nelson JA (June 2012). "Invention of the N-localizer for stereotactic neurosurgery and its use in the Brown-Roberts-Wells stereotactic frame". *Neurosurgery* **70** (2 Supplement Operative): 173–176. doi:10.1227/NEU.0b013e318246a4f7. PMID 22186842.

[35] Heilbrun MP; Roberts TS; Apuzzo ML; Wells TH Jr; et al. (August 1983). "Preliminary experience with Brown-Roberts-Wells (BRW) computerized tomography stereotaxic guidance system". *Journal of Neurosurgery* **59** (2): 217–222. doi:10.3171/jns.1 983.59.2.0217.PMID6345727.

[36] Thomas DG; Anderson RE; du Boulay GH (January 1984). "CT-guided stereotactic neurosurgery: experience in 24 cases with a new stereotactic system". *Journal of Neurology, Neurosurgery & Psychiatry* **47** (1): 9–16. doi:10.1136/jnnp.47.1.9. PMC 1027634. PMID 6363629.

[37] Leksell L; Leksell D; Schwebel J (January 1985). "Stereotaxis and nuclear magnetic resonance". *Journal of Neurology, Neurosurgery & Psychiatry* **48** (1): 14–18. doi:10.1136/jnnp.48.1.14. PMC 1028176. PMID 3882889.

[38] Thomas DG; Davis CH; Ingram S; Olney JS; et al. (January 1986). "Stereotaxic biopsy of the brain under MR imaging control". *AJNR American Journal of Neuroradiology* **7** (1): 161–163. PMID 3082131.

[39] Heilbrun MP; Sunderland PM; McDonald PR; Wells TH Jr; et al. (1987). "Brown-Roberts-Wells stereotactic frame modifications to accomplish magnetic resonance imaging guidance in three planes". *Applied Neurophysiology* **50** (1–6): 143–152. doi:10.1159/000100700. PMID 3329837.

[40] Maciunas RJ; Kessler RM; Maurer C; Mandava V; et al. (1992). "Positron emission tomography imaging-directed stereotactic neurosurgery". *Stereotactic and Functional Neurosurgery* **58** (1–4): 134–140. doi:10.1159/000098986. PMID 1439330.

[41] Bracco Diagnostics, CardioGen-82, 2000

[42] Young H; Baum R; Cremerius U; Herholz K; et al. (1999). "Measurement of clinical and subclinical tumour response using [18F]-fluorodeoxyglucose and positron emission tomography: review and 1999 EORTC recommendations". *European Journal of Cancer* **35** (13): 1773–1782. doi:10.1016/S0959-8049(99)00229-4. PMID 10673991.

[43] Technology | July 2003: Trends in MRI | Medical Imaging

[44] Michael Phelps Talk on PET Scans

[45] Vardi, Y.; L. A. Shepp; L. Kaufman (1985). "A statistical model for positron emission tomography". *Journal of the American Statistical Association* **80** (389): 8–37. doi:10.1080/01621459.1985.10477119.

[46] Willett, R.; Z. Harmany; R. Marcia (2010). "Poisson Image Reconstruction with Total Variation Regularization". *Accepted to IEEE International Conference on Image Processing (ICIP).*

[47] Harmany, Z.; R. Marcia; R. Willett (2010). "Sparsity-regularized Photon-limited Imaging". *International Symposium on Biomedical Imaging (ISBI).*

[48] Harmany, Z.; R. Marcia; R. Willett (2010). "SPIRAL out of Convexity: Sparsity-regularized Algorithms for Photon-limited Imaging". *SPIE Electronic Imaging.*

[49] Ter-Pogossian MM; Phelps ME; Hoffman EJ; Mullani NA (1975). "A positron-emission transaxial tomograph for nuclear imaging (PET)". *Radiology* **114** (1): 89–98. doi:10.1148/114.1.89. OSTI 4251398. PMID 1208874.

[50] Phelps ME; Hoffman EJ; Mullani NA; Ter-Pogossian MM (March 1, 1975). "Application of annihilation coincidence detection to transaxial reconstruction tomography". *Journal of Nuclear Medicine* **16** (3): 210–224. PMID 1113170.

[51] Sweet, W.H.; G.L. Brownell (1953). "Localization of brain tumors with positron emitters". *Nucleonics* **11**: 40–45.

[52] *A Vital Legacy: Biological and Environmental Research in the Atomic Age,* U.S. Department of Energy, The Office of Biological and Environmental Research, September 2010, p 25–26

[53] Ido T; Wan CN; Casella V; Fowler JS; et al. (1978). "Labeled 2-deoxy-D-glucose analogs. 18F-labeled 2-deoxy-2-fluoro-D-glucose, 2-deoxy-2-fluoro-D-mannose and 14C-2-deoxy-2-fluoro-D-glucose". *Journal of Labelled Compounds and Radiopharmaceuticals* **14** (2): 175–183. doi:10.1002/jlcr.2580140204.

[54] BROWNELL G.L., Dave Marcum, B. HOOP JR., and D.E. BOHNING, "Quantitative dynamic studies using short-lived radioisotopes and positron detection" in Proceedings of the Symposium on Dynamic Studies with Radioisotopes in Medicine, Rotterdam. August 31–September 4, 1945. IAEA. Vienna. 194824. pp. 161–172.

[55] ROBERTSON J.S., MARR R.B., ROSENBLUM M., RADEKA V., and YAMAMOTO Y.L., ``32-Crystal positron transverse section detector, *in Tomographic Imaging in Nuclear Medicine, Freedman GS, Editor. 1983, The Society of Nuclear Medicine: New York. pp. 142–153.*

[56] CHO, Z. H., ERIKSSON L., and CHAN J.K., ``A circular ring transverse axial positron camera *in Reconstruction Tomography in Diagnostic Radiology and Nuclear Medicine, Ed. Ter-Pogossian MM., University Park Press: Baltimore, 1975.*

[57] "PET Scan: PET/CT History". Petscaninfo.com. Retrieved 2012-08-13.

[58] Ontario PET Steering Committee (August 31, 2008), *PET SCAN PRIMER, A Guide to the Implementation of Positron Emission Tomography Imaging in Ontario, Executive Summary*, pp. iii

4.9 External links

- PET Images Search MedPix(r)

- Seeing is believing: In vivo functional real-time imaging of transplanted islets using positron emission tomography (PET) (a protocol), Nature Protocols, from Nature Medicine 12, 1423–1428 (2006).

- The nuclear medicine and molecular medicine podcast—Podcast

- Positron Emission Particle Tracking (PEPT)—engineering analysis tool based on PET that is able to track single particles in 3D within mixing systems or fluidised beds. Developed at the University of Birmingham, UK.

- CMS coverage of PET scans

- PET-CT atlas Harvard Medical School

- National Isotope Development Center—U.S. government source of radionuclides including those for PET—produ research, development, distribution, and information

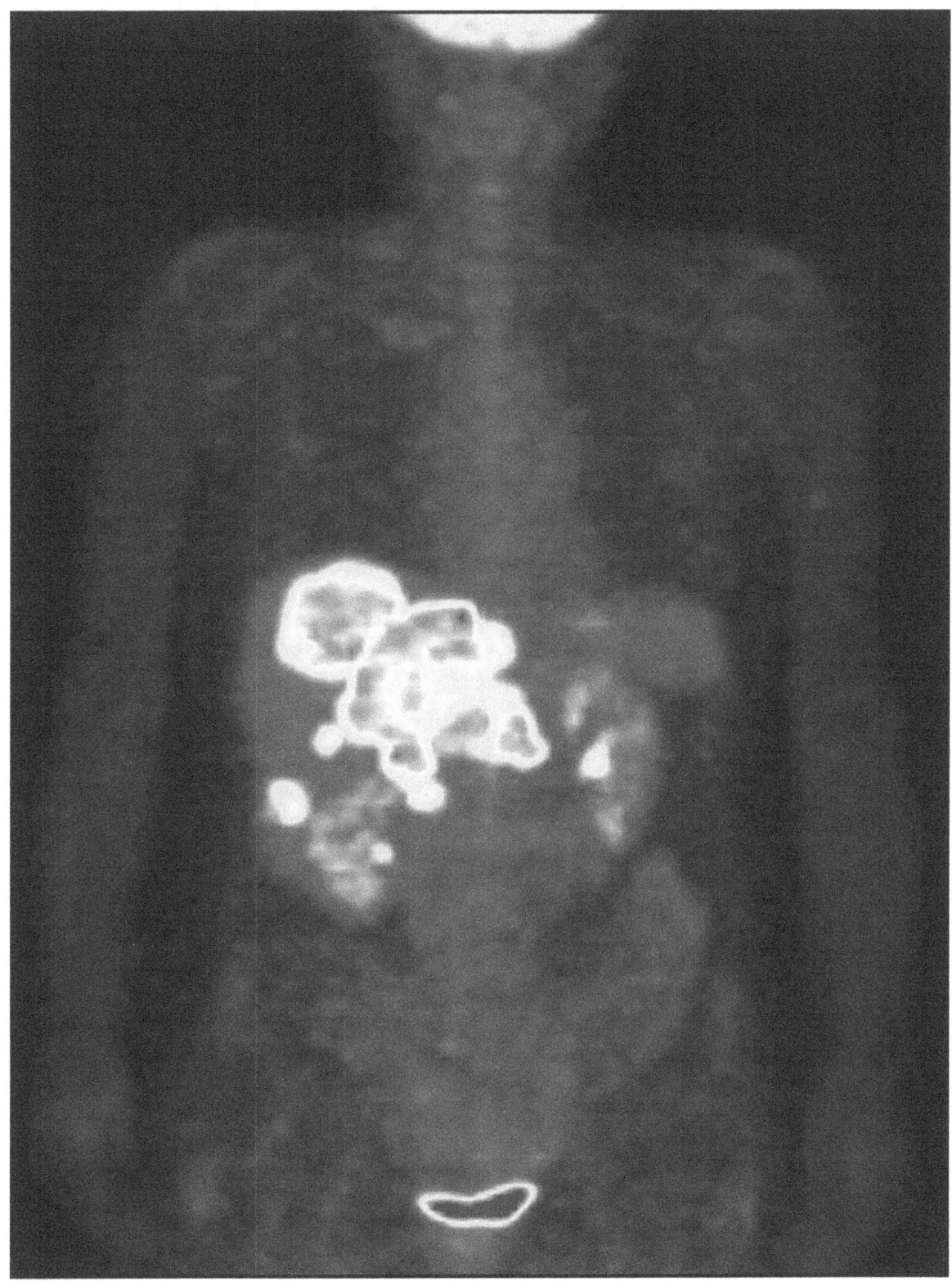

Whole-body PET scan using ^{18}F-FDG

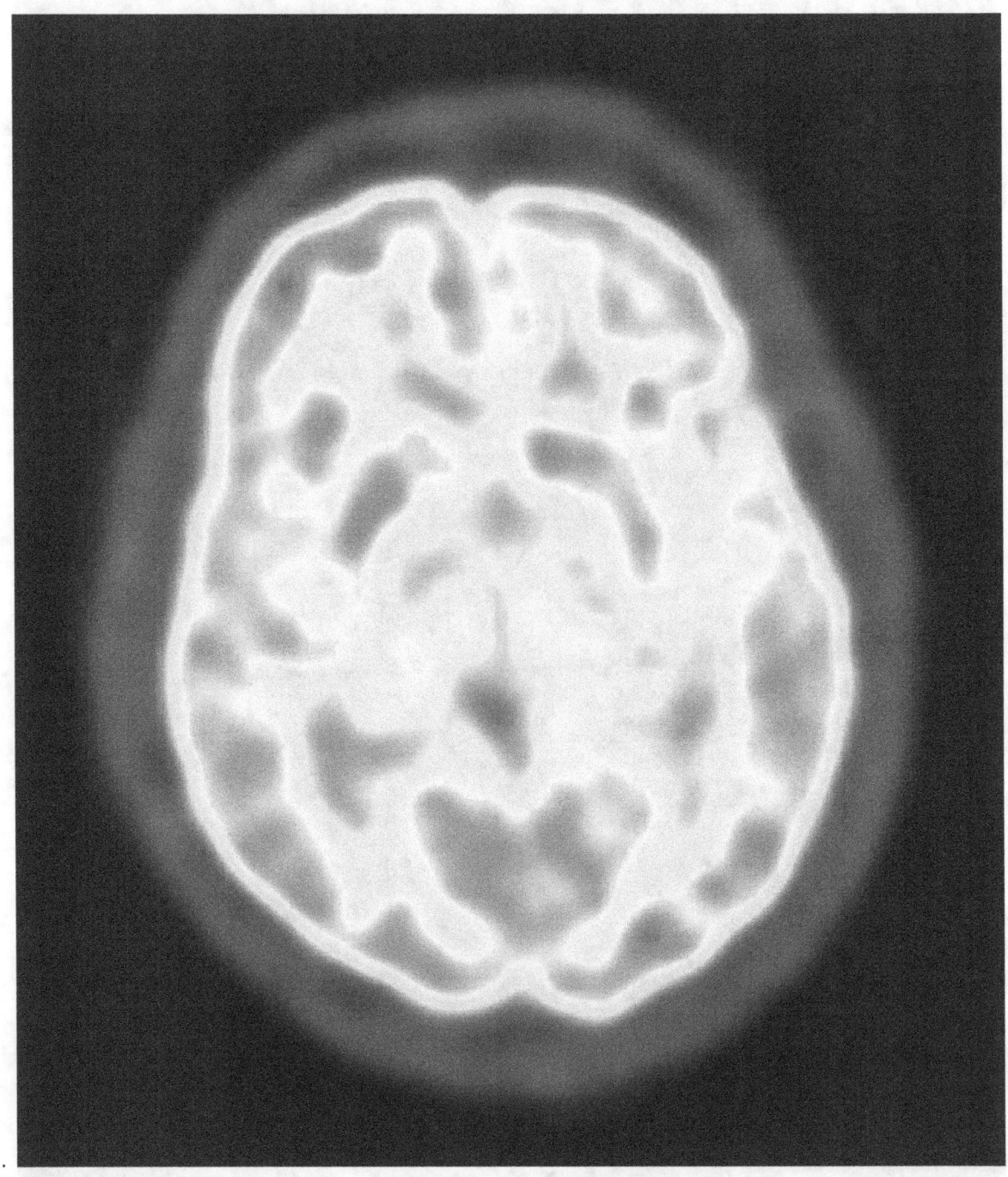

1.
PET scan of the human brain.

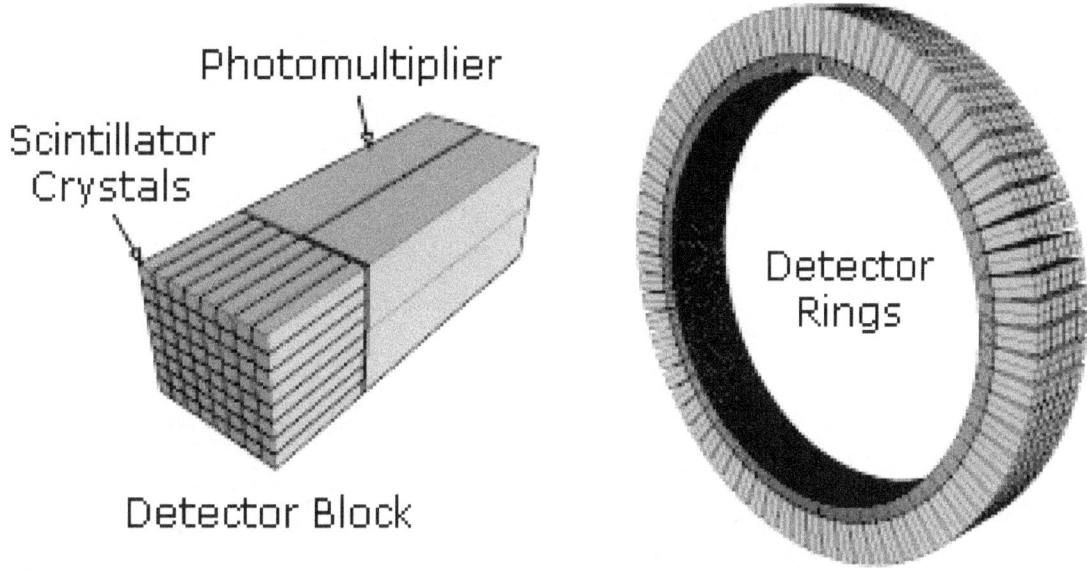

Schematic view of a detector block and ring of a PET scanner

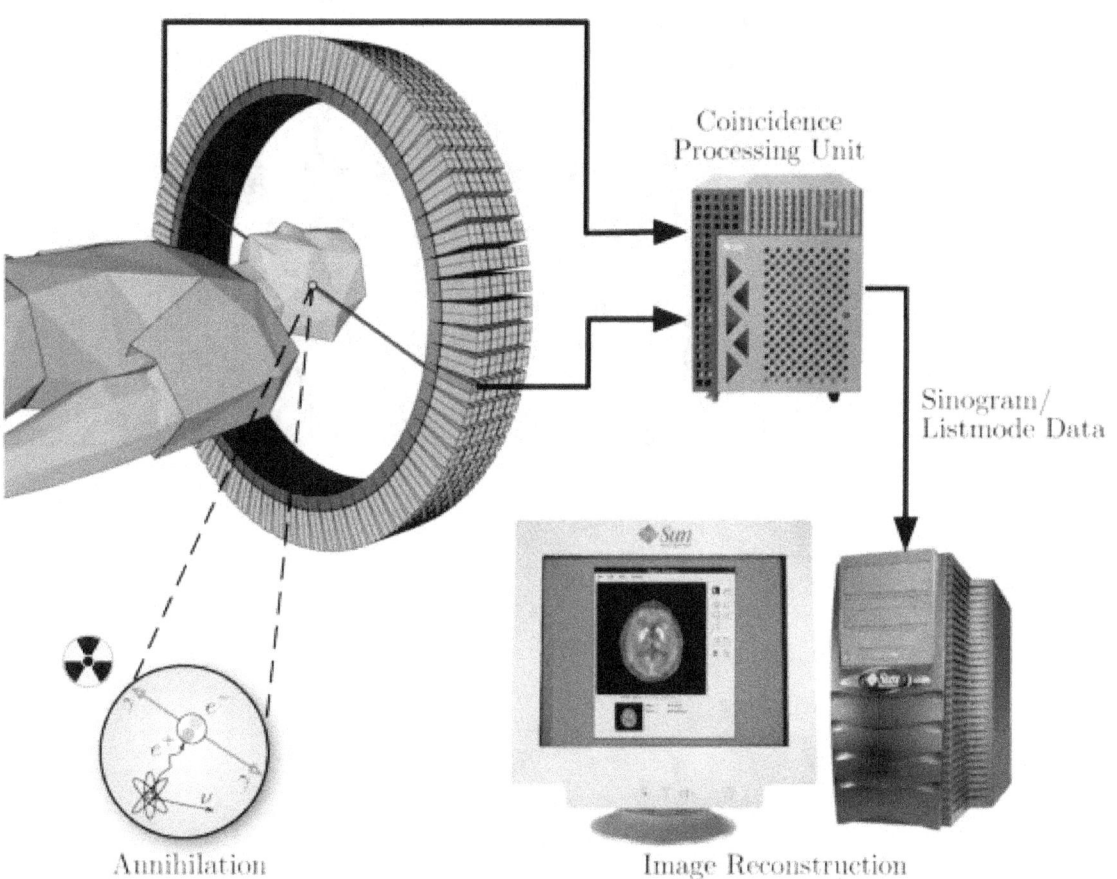

Schema of a PET acquisition process

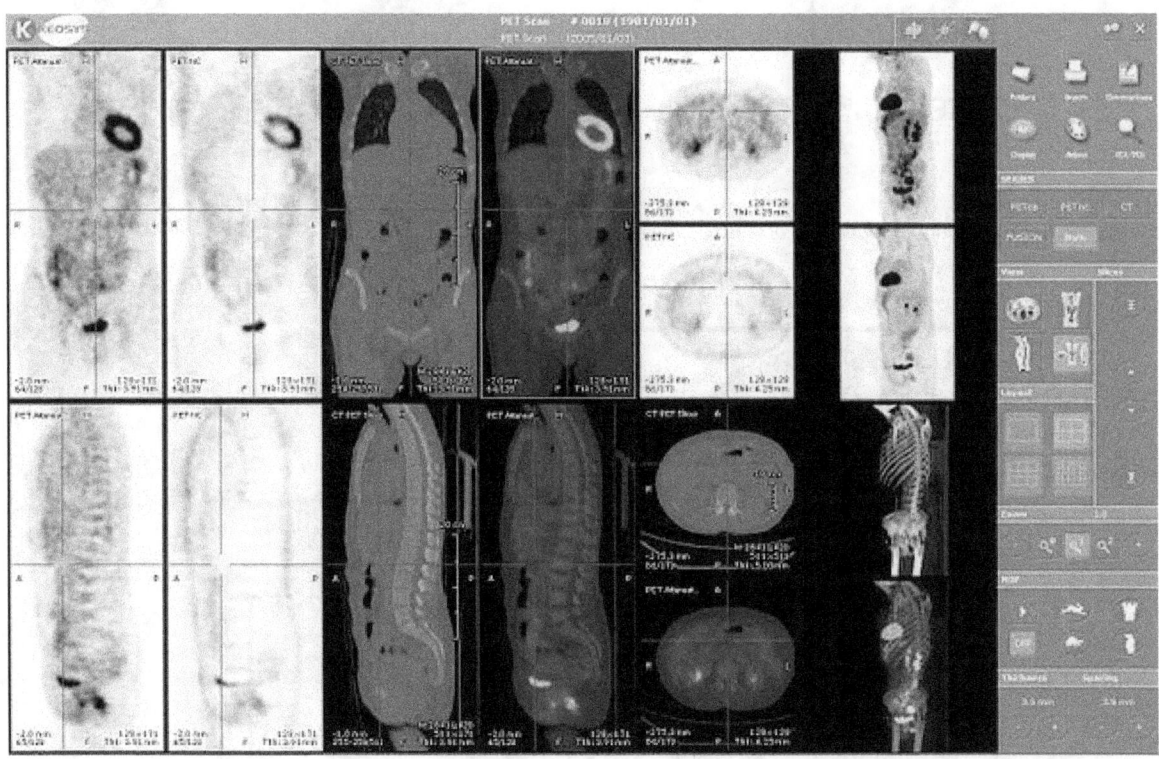

Complete body PET-CT fusion image

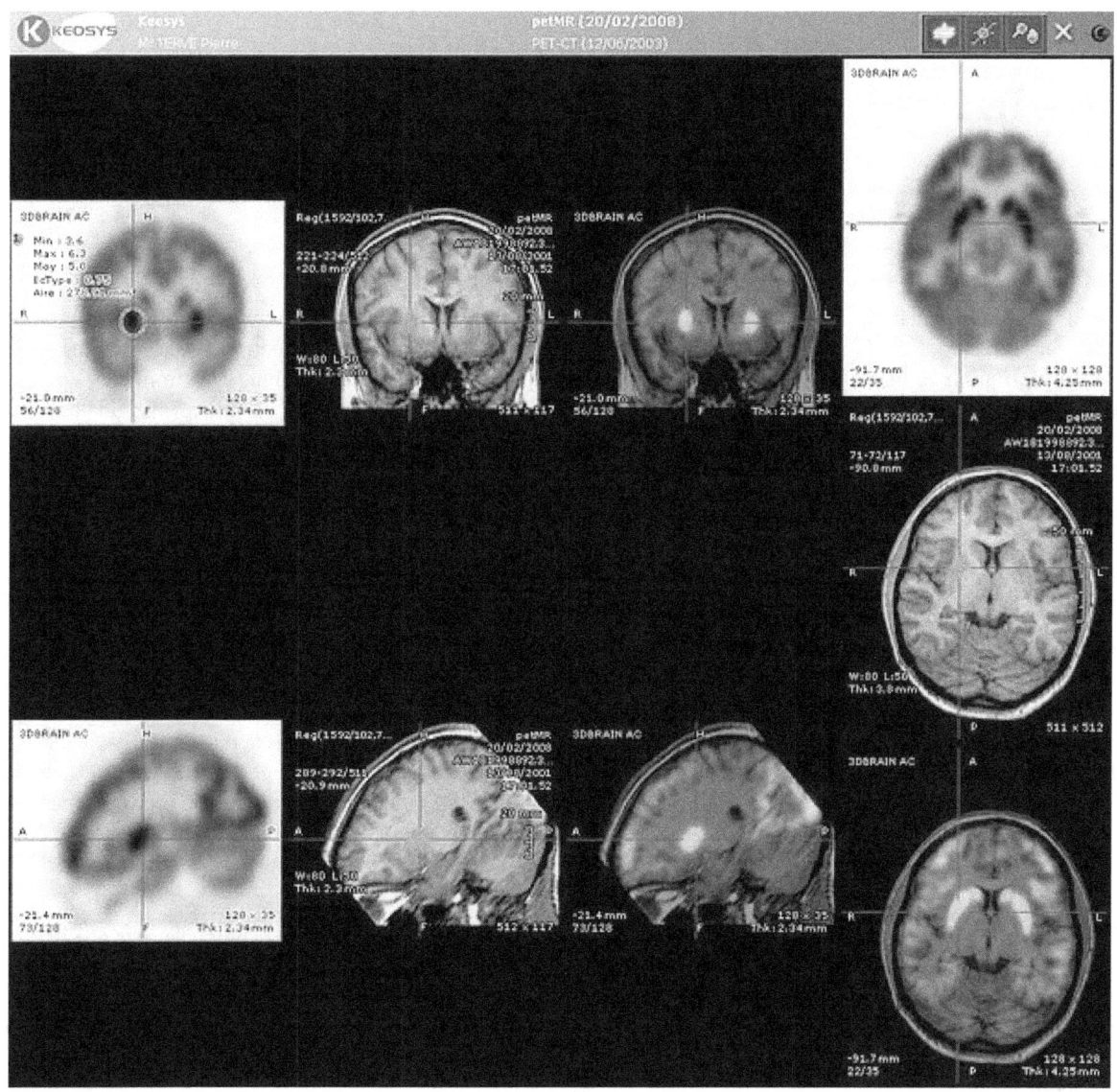

Brain PET-MRI fusion image

Chapter 5

Cosmic ray

Cosmic rays are immensely high-energy radiation, mainly originating outside the Solar System. [1] They may produce showers of secondary particles that penetrate and impact the Earth's atmosphere and sometimes even reach the surface. Composed primarily of high-energy protons and atomic nuclei, they are of mysterious origin. Data from the *Fermi* space telescope (2013)[2] has been interpreted as evidence that a significant fraction of primary cosmic rays originate from the supernovae of massive stars.[3] However, this is not thought to be their only source. Active galactic nuclei probably also produce cosmic rays.

The term *ray* is a historical accident, as cosmic rays were at first, and wrongly, thought to be mostly electromagnetic radiation. In common scientific usage[4] high-energy particles with intrinsic mass are known as "cosmic" rays, and photons, which are quanta of electromagnetic radiation (and so have no intrinsic mass) are known by their common names, such as "gamma rays" or "X-rays", depending on their origin.

Cosmic rays attract great interest practically, due to the damage they inflict on microelectronics and life outside the protection of an atmosphere and magnetic field, and scientifically, because the energies of the most energetic ultra-high-energy cosmic rays (UHECRs) have been observed to approach 3×10^{20} eV,[5] about 40 million times the energy of particles accelerated by the Large Hadron Collider.[6] One can show that such enormous energies might be achieved by means of the Centrifugal mechanism of acceleration in Active galactic nuclei. At 50 J,[7] the highest-energy ultra-high-energy cosmic rays have energies comparable to the kinetic energy of a 90-kilometre-per-hour (56 mph) baseball. As a result of these discoveries, there has been interest in investigating cosmic rays of even greater energies.[8] Most cosmic rays, however, do not have such extreme energies; the energy distribution of cosmic rays peaks at 0.3 gigaelectronvolts $(4.8 \times 10^{-11}$ J).[9]

Of primary cosmic rays, which originate outside of Earth's atmosphere, about 99% are the nuclei (stripped of their electron shells) of well-known atoms, and about 1% are solitary electrons (similar to beta particles). Of the nuclei, about 90% are simple protons, i. e. hydrogen nuclei; 9% are alpha particles, and 1% are the nuclei of heavier elements, called HZE ions.[10] A very small fraction are stable particles of antimatter, such as positrons or antiprotons. The precise nature of this remaining fraction is an area of active research. An active search from Earth orbit for anti-alpha particles has failed to detect them.

5.1 History

After the discovery of radioactivity by Henri Becquerel and Marie Curie in 1896, it was generally believed that atmospheric electricity, ionization of the air, was caused only by radiation from radioactive elements in the ground or the radioactive gases or isotopes of radon they produce.[11] Measurements of ionization rates at increasing heights above the ground during the decade from 1900 to 1910 showed a decrease that could be explained as due to absorption of the ionizing radiation by the intervening air.[12]

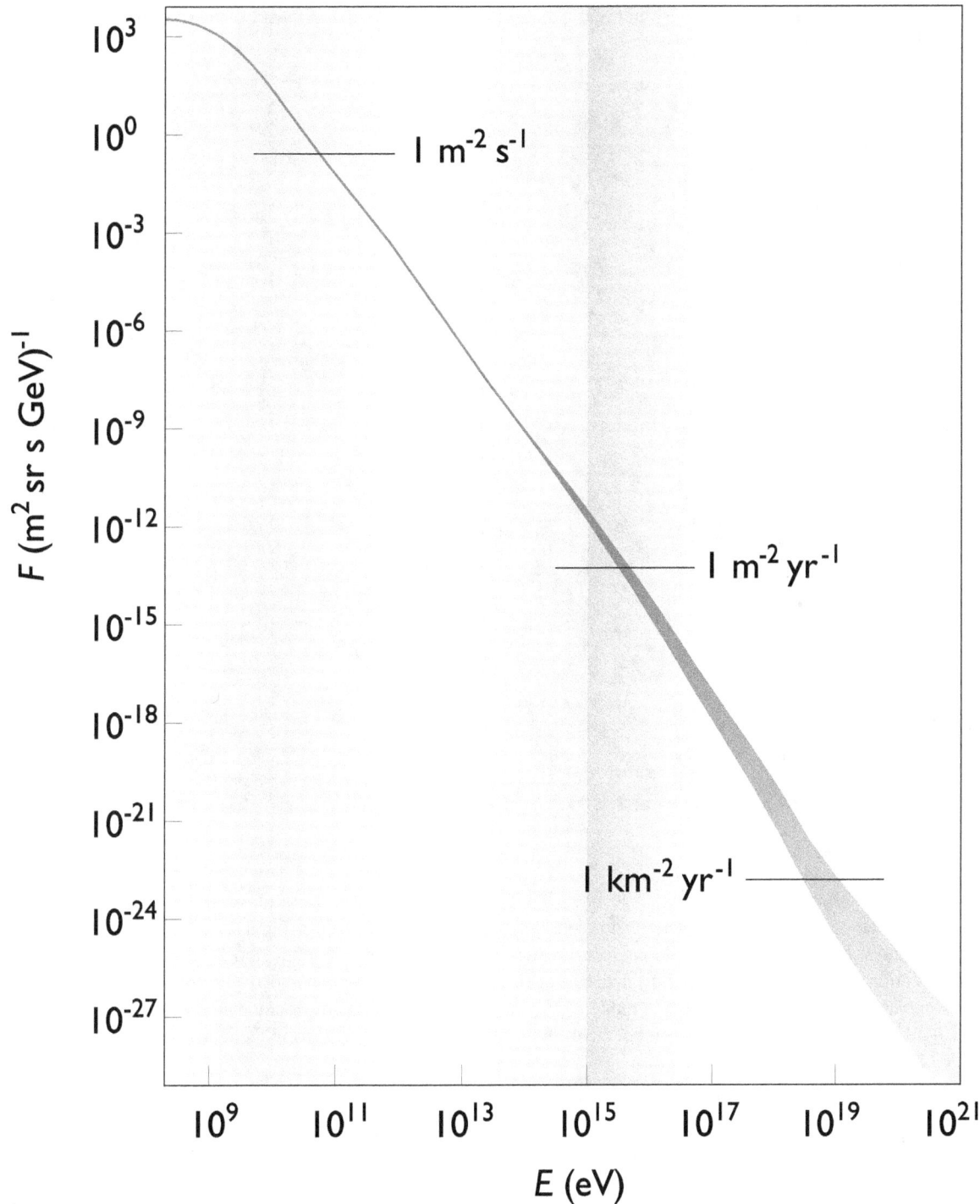

Cosmic ray flux versus particle energy

5.1.1 Discovery

In 1909 Theodor Wulf developed an electrometer, a device to measure the rate of ion production inside a hermetically sealed container, and used it to show higher levels of radiation at the top of the Eiffel Tower than at its base. However, his paper published in *Physikalische Zeitschrift* was not widely accepted. In 1911 Domenico Pacini observed simultaneous variations of the rate of ionization over a lake, over the sea, and at a depth of 3 meters from the surface. Pacini concluded from the decrease of radioactivity underwater that a certain part of the ionization must be due to sources other than the radioactivity of the Earth.[13]

Pacini makes a measurement in 1910.

In 1912, Victor Hess carried three enhanced-accuracy Wulf electrometers[14] to an altitude of 5300 meters in a free balloon flight. He found the ionization rate increased approximately fourfold over the rate at ground level.[14] Hess ruled out the Sun as the radiation's source by making a balloon ascent during a near-total eclipse. With the moon blocking much of the Sun's visible radiation, Hess still measured rising radiation at rising altitudes.[14] He concluded "The results of my observation are best explained by the assumption that a radiation of very great penetrating power enters our atmosphere from above." In 1913–1914, Werner Kolhörster confirmed Victor Hess' earlier results by measuring the increased ionization rate at an altitude of 9 km.

Hess received the Nobel Prize in Physics in 1936 for his discovery.[15][16]

The Hess balloon flight took place on 7 August 1912. By sheer coincidence, exactly 100 years later on 7 August 2012, the Mars Science Laboratory rover used its Radiation Assessment Detector (RAD) instrument to begin measuring the radiation levels on another planet for the first time. On 31 May 2013, NASA scientists reported that a possible manned mission to Mars may involve a greater radiation risk than previously believed, based on the amount of energetic particle radiation detected by the RAD on the Mars Science Laboratory while traveling from the Earth to Mars in 2011–2012.[17][18][19]

5.1.2 Identification

In the 1920s the term "cosmic rays" was coined by Robert Millikan who made measurements of ionization due to cosmic rays from deep under water to high altitudes and around the globe. Millikan believed that his measurements proved that the primary cosmic rays were gamma rays, i.e., energetic photons. And he proposed a theory that they were produced in interstellar space as by-products of the fusion of hydrogen atoms into the heavier elements, and that secondary electrons

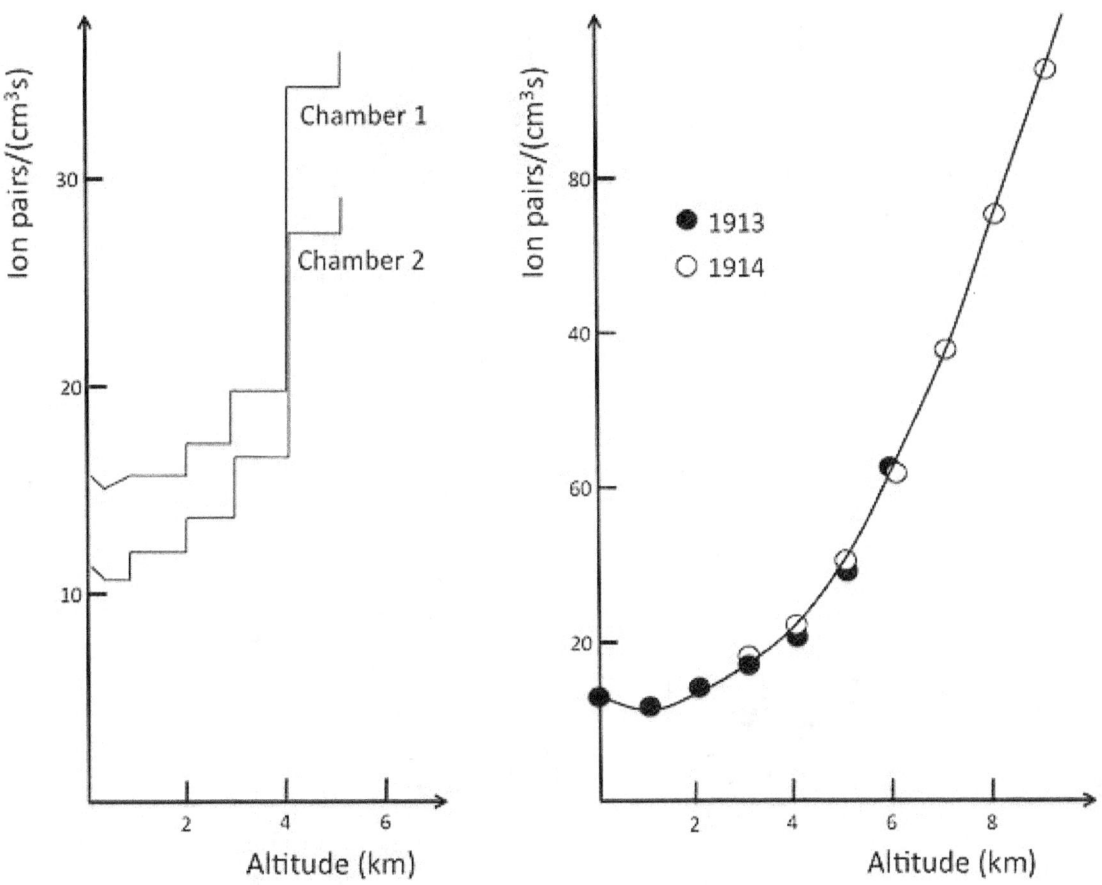

Increase of ionization with altitude as measured by Hess in 1912 (left) and by Kolhörster (right)

were produced in the atmosphere by Compton scattering of gamma rays. But then, in 1927, J. Clay found evidence,[20] later confirmed in many experiments, of a variation of cosmic ray intensity with latitude, which indicated that the primary cosmic rays are deflected by the geomagnetic field and must therefore be charged particles, not photons. In 1929, Bothe and Kolhörster discovered charged cosmic-ray particles that could penetrate 4.1 cm of gold.[21] Charged particles of such high energy could not possibly be produced by photons from Millikan's proposed interstellar fusion process.

In 1930, Bruno Rossi predicted a difference between the intensities of cosmic rays arriving from the east and the west that depends upon the charge of the primary particles – the so-called "east-west effect."[22] Three independent experiments[23][24][25] found that the intensity is, in fact, greater from the west, proving that most primaries are positive. During the years from 1930 to 1945, a wide variety of investigations confirmed that the primary cosmic rays are mostly protons, and the secondary radiation produced in the atmosphere is primarily electrons, photons and muons. In 1948, observations with nuclear emulsions carried by balloons to near the top of the atmosphere showed that approximately 10% of the primaries are helium nuclei (alpha particles) and 1% are heavier nuclei of the elements such as carbon, iron, and lead.[26][27]

During a test of his equipment for measuring the east-west effect, Rossi observed that the rate of near-simultaneous discharges of two widely separated Geiger counters was larger than the expected accidental rate. In his report on the experiment, Rossi wrote "... it seems that once in a while the recording equipment is struck by very extensive showers of particles, which causes coincidences between the counters, even placed at large distances from one another."[25] In 1937 Pierre Auger, unaware of Rossi's earlier report, detected the same phenomenon and investigated it in some detail. He concluded that high-energy primary cosmic-ray particles interact with air nuclei high in the atmosphere, initiating a cascade of secondary interactions that ultimately yield a shower of electrons, and photons that reach ground level.[28]

Soviet physicist Sergey Vernov was the first to use radiosondes to perform cosmic ray readings with an instrument carried

Hess lands after his balloon flight in 1912.

to high altitude by a balloon. On 1 April 1935, he took measurements at heights up to 13.6 kilometers using a pair of Geiger counters in an anti-coincidence circuit to avoid counting secondary ray showers.[29][30]

Homi J. Bhabha derived an expression for the probability of scattering positrons by electrons, a process now known as Bhabha scattering. His classic paper, jointly with Walter Heitler, published in 1937 described how primary cosmic rays from space interact with the upper atmosphere to produce particles observed at the ground level. Bhabha and Heitler explained the cosmic ray shower formation by the cascade production of gamma rays and positive and negative electron pairs.

5.1.3   Energy distribution

Measurements of the energy and arrival directions of the ultra-high energy primary cosmic rays by the techniques of "density sampling" and "fast timing" of extensive air showers were first carried out in 1954 by members of the Rossi Cosmic Ray Group at the Massachusetts Institute of Technology.[31] The experiment employed eleven scintillation detectors arranged within a circle 460 meters in diameter on the grounds of the Agassiz Station of the Harvard College Observatory. From that work, and from many other experiments carried out all over the world, the energy spectrum of the primary cosmic rays is now known to extend beyond 10^{20} eV. A huge air shower experiment called the Auger Project is currently operated at a site on the pampas of Argentina by an international consortium of physicists, led by James Cronin, winner of the 1980 Nobel Prize in Physics from the University of Chicago, and Alan Watson of the University of Leeds. Their aim is to explore the properties and arrival directions of the very highest-energy primary cosmic rays.[32] The results are expected to have important implications for particle physics and cosmology, due to a theoretical Greisen–Zatsepin–Kuzmin limit to the energies of cosmic rays from long distances (about 160 million light years) which occurs above 10^{20} eV because of interactions with the remnant photons from the Big Bang origin of the universe.

High-energy gamma rays (>50 MeV photons) were finally discovered in the primary cosmic radiation by an MIT experiment carried on the OSO-3 satellite in 1967.[33] Components of both galactic and extra-galactic origins were separately identified at intensities much less than 1% of the primary charged particles. Since then, numerous satellite gamma-ray observatories have mapped the gamma-ray sky. The most recent is the Fermi Observatory, which has produced a map showing a narrow band of gamma ray intensity produced in discrete and diffuse sources in our galaxy, and numerous point-like extra-galactic sources distributed over the celestial sphere.

5.2   Sources of cosmic rays

Early speculation on the sources of cosmic rays included a 1934 proposal by Baade and Zwicky suggesting cosmic rays originated from supernovae.[34] A 1948 proposal by Horace W. Babcock suggested that magnetic variable stars could be a source of cosmic rays.[35] Subsequently in 1951, Y. Sekido *et al.* identified the Crab Nebula as a source of cosmic rays.[36] Since then, a wide variety of potential sources for cosmic rays began to surface, including supernovae, active galactic nuclei, quasars, and gamma-ray bursts.[37]

Later experiments have helped to identify the sources of cosmic rays with greater certainty. In 2009, a paper presented at the International Cosmic Ray Conference (ICRC) by scientists at the Pierre Auger Observatory showed ultra-high energy cosmic rays (UHECRs) originating from a location in the sky very close to the radio galaxy Centaurus A, although the authors specifically stated that further investigation would be required to confirm Cen A as a source of cosmic rays.[38] However, no correlation was found between the incidence of gamma-ray bursts and cosmic rays, causing the authors to set upper limits as low as 3.4×10^{-6} erg cm^{-2} on the flux of 1 GeV-1 TeV cosmic rays from gamma-ray bursts.[39]

In 2009, supernovae were said to have been "pinned down" as a source of cosmic rays, a discovery made by a group using data from the Very Large Telescope.[40] This analysis, however, was disputed in 2011 with data from PAMELA, which revealed that "spectral shapes of [hydrogen and helium nuclei] are different and cannot be described well by a single power law", suggesting a more complex process of cosmic ray formation.[41] In February 2013, though, research analyzing data from *Fermi* revealed through an observation of neutral pion decay that supernovae were indeed a source of cosmic rays, with each explosion producing roughly 3×10^{42} - 3×10^{43} J of cosmic rays.[2][3] However, supernovae do not produce all cosmic rays, and the proportion of cosmic rays that they do produce is a question which cannot be answered without further study.[42]

5.3   Types

Cosmic rays can be divided into two types, **Galactic Cosmic Rays** ("GCR"), high energy particles originating outside the solar system, and **Solar energetic particles**, high energy particles (predominantly protons) emitted by the sun, primarily in solar particle events. However, the term "cosmic ray" is often used to refer to only the GCR flux. Despite the nomenclature "galactic", GRCs may originate within or outside the galaxy (as discussed in the source section above).

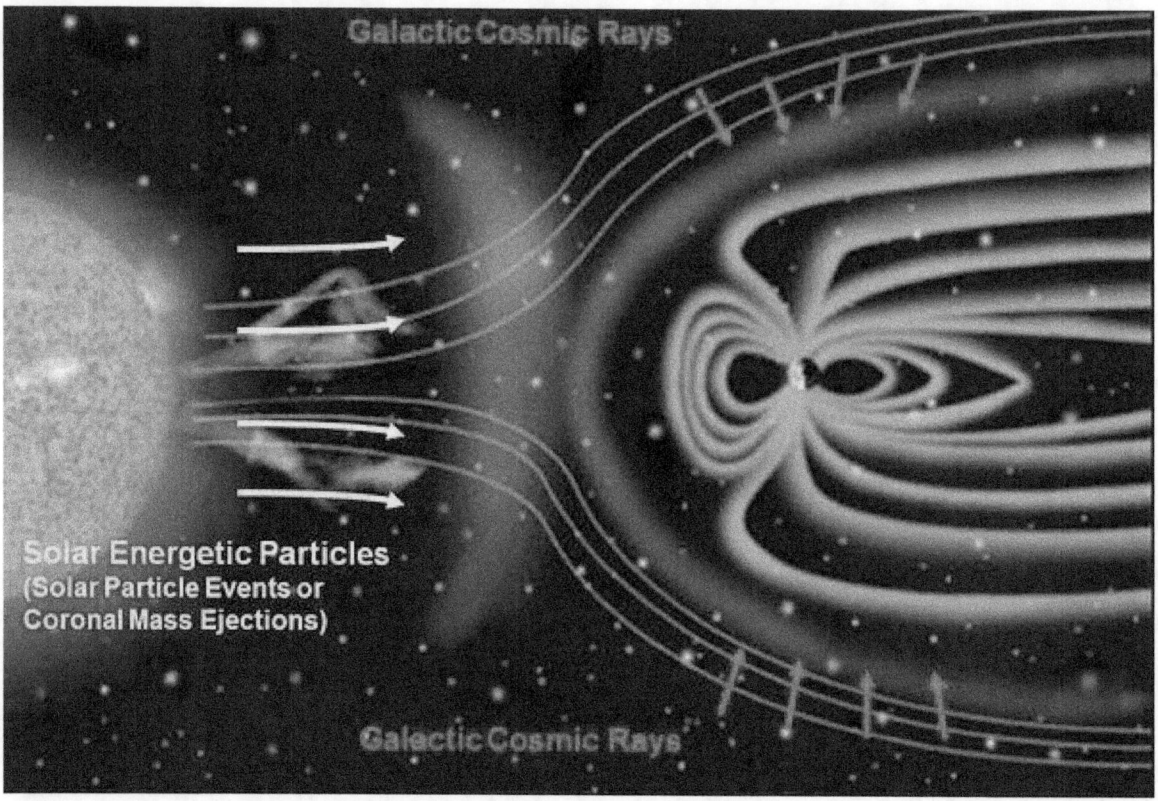

Sources of ionizing radiation in interplanetary space.

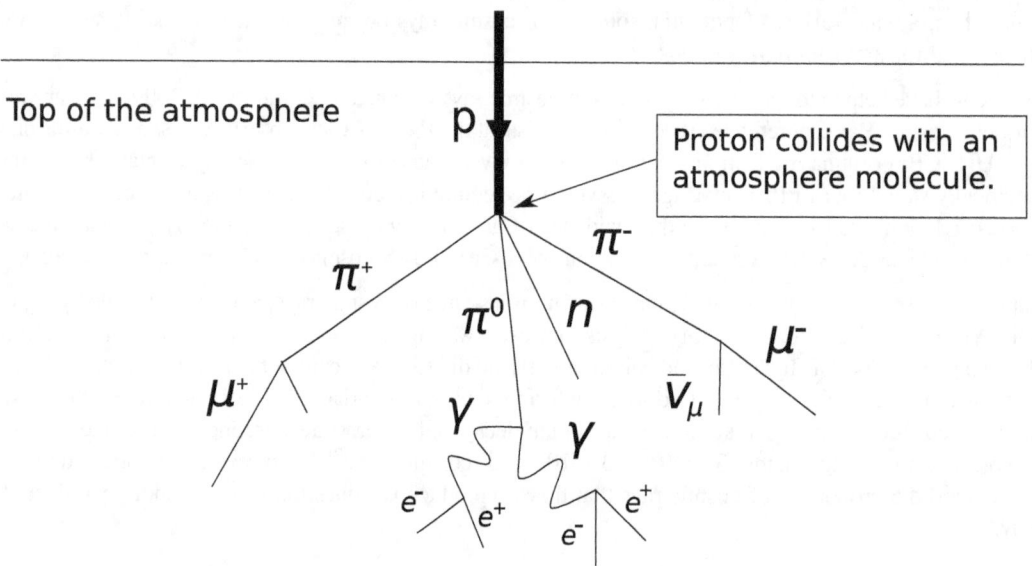

Primary cosmic particle collides with a molecule of atmosphere.

Cosmic rays originate as primary cosmic rays, which are those originally produced in various astrophysical processes. Primary cosmic rays are composed primarily of protons and alpha particles (99%), with a small amount of heavier nuclei (~1%) and an extremely minute proportion of positrons and antiprotons.[10] Secondary cosmic rays, caused by a decay of primary cosmic rays as they impact an atmosphere, include neutrons, pions, positrons, and muons. Of these four, the latter three were first detected in cosmic rays.

5.3.1 Primary cosmic rays

Primary cosmic rays primarily originate from outside the Solar System and sometimes even the Milky Way. When they interact with Earth's atmosphere, they are converted to secondary particles. The mass ratio of helium to hydrogen nuclei, 28%, is similar to the primordial elemental abundance ratio of these elements, 24%.[43] The remaining fraction is made up of the other heavier nuclei that are typical nucleosynthesis end products, primarily lithium, beryllium, and boron. These nuclei appear in cosmic rays in much greater abundance (~1%) than in the solar atmosphere, where they are only about 10^{-11} as abundant as helium. Cosmic rays made up of charged nuclei heavier than helium are called HZE ions. Due to the high charge and heavy nature of HZE ions, their contribution to an astronaut's radiation dose in space is significant even though they are relatively scarce.

This abundance difference is a result of the way secondary cosmic rays are formed. Carbon and oxygen nuclei collide with interstellar matter to form lithium, beryllium and boron in a process termed cosmic ray spallation. Spallation is also responsible for the abundances of scandium, titanium, vanadium, and manganese ions in cosmic rays produced by collisions of iron and nickel nuclei with interstellar matter.[44]

Primary cosmic ray antimatter

See also: Alpha Magnetic Spectrometer

Satellite experiments have found evidence of positrons and a few antiprotons in primary cosmic rays, amounting to less than 1% of the particles in primary cosmic rays. These do not appear to be the products of large amounts of antimatter from the Big Bang, or indeed complex antimatter in the universe. Rather, they appear to consist of only these two elementary particles, newly made in energetic processes.

Preliminary results from the presently operating Alpha Magnetic Spectrometer (*AMS-02*) on board the International Space Station show that positrons in the cosmic rays arrive with no directionality, and with energies that range from 10 GeV to 250 GeV. In September, 2014, new results with almost twice as much data were presented in a talk at CERN and published in Physical Review Letters.[45][46] A new measurement of positron fraction up to 500 GeV was reported, showing that positron fraction peaks at a maximum of about 16% of total electron+positron events, around an energy of 275 ± 32 GeV. At higher energies, up to 500 GeV, the ratio of positrons to electrons begins to fall again. The absolute flux of positrons also begins to fall before 500 GeV, but peaks at energies far higher than electron energies, which peak about 10 GeV.[47] These results on interpretation have been suggested to be due to positron production in annihilation events of massive dark matter particles.[48]

Cosmic ray antiprotons also have a much higher energy than their normal-matter counterparts (protons). They arrive at Earth with a characteristic energy maximum of 2 GeV, indicating their production in a fundamentally different process from cosmic ray protons, which on average have only one-sixth of the energy.[49]

There is no evidence of complex antimatter atomic nuclei, such as antihelium nuclei (i.e., anti-alpha particles), in cosmic rays. These are actively being searched for. A prototype of the *AMS-02* designated *AMS-01*, was flown into space aboard the Space Shuttle *Discovery* on STS-91 in June 1998. By not detecting any antihelium at all, the *AMS-01* established an upper limit of 1.1×10^{-6} for the antihelium to helium flux ratio.[50]

The moon in cosmic rays

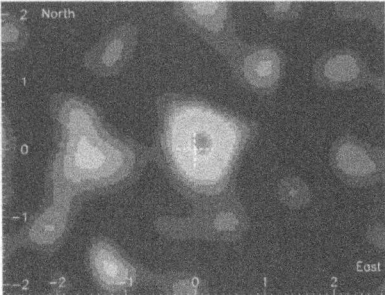

The Moon's cosmic ray shadow, as seen in secondary muons detected 700 m below ground, at the Soudan 2 detector

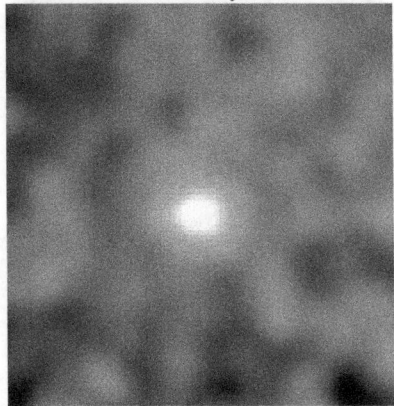

The moon as seen by the Compton Gamma Ray Observatory, in gamma rays with energies greater than 20 MeV. These are produced by cosmic ray bombardment on its surface.[51]

5.3.2 Secondary cosmic rays

When cosmic rays enter the Earth's atmosphere they collide with atoms and molecules, mainly oxygen and nitrogen. The interaction produces a cascade of lighter particles, a so-called air shower secondary radiation that rains down, including x-rays, muons, protons, alpha particles, pions, electrons, and neutrons.[52] All of the produced particles stay within about one degree of the primary particle's path.

Typical particles produced in such collisions are neutrons and charged mesons such as positive or negative pions and kaons. Some of these subsequently decay into muons, which are able to reach the surface of the Earth, and even penetrate for some distance into shallow mines. The muons can be easily detected by many types of particle detectors, such as cloud chambers, bubble chambers or scintillation detectors. The observation of a secondary shower of particles in multiple detectors at the same time is an indication that all of the particles came from that event.

Cosmic rays impacting other planetary bodies in the Solar System are detected indirectly by observing high energy gamma ray emissions by gamma-ray telescope. These are distinguished from radioactive decay processes by their higher energies above about 10 MeV.

5.3.3 Cosmic-ray flux

The flux of incoming cosmic rays at the upper atmosphere is dependent on the solar wind, the Earth's magnetic field, and the energy of the cosmic rays. At distances of ~94 AU from the Sun, the solar wind undergoes a transition, called the termination shock, from supersonic to subsonic speeds. The region between the termination shock and the heliopause acts as a barrier to cosmic rays, decreasing the flux at lower energies (≤ 1 GeV) by about 90%. However, the strength of the solar wind is not constant, and hence it has been observed that cosmic ray flux is correlated with solar activity.

In addition, the Earth's magnetic field acts to deflect cosmic rays from its surface, giving rise to the observation that the flux is apparently dependent on latitude, longitude, and azimuth angle. The magnetic field lines deflect the cosmic rays

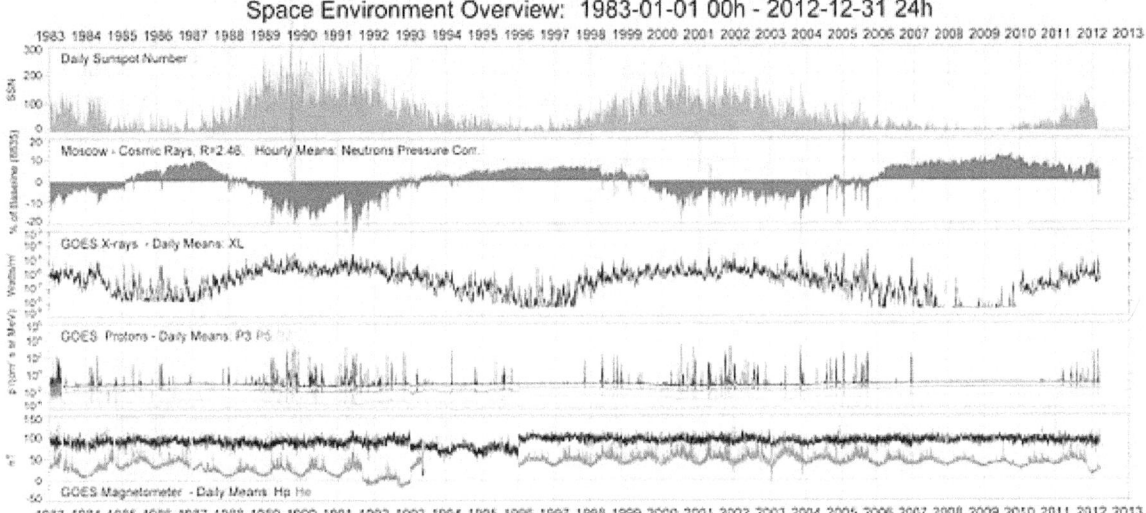

An overview of the space environment shows the relationship between the solar activity and galactic cosmic rays.[53]

towards the poles, giving rise to the aurorae.

The combined effects of all of the factors mentioned contribute to the flux of cosmic rays at Earth's surface. The following table of participial frequencies reach the planet[54] and are inferred from lower energy radiation reaching the ground[55]

In the past, it was believed that the cosmic ray flux remained fairly constant over time. However, recent research suggests 1.5 to 2-fold millennium-timescale changes in the cosmic ray flux in the past forty thousand years.[56]

The magnitude of the energy of cosmic ray flux in interstellar space is very comparable to that of other deep space energies: cosmic ray energy density averages about one electron-volt per cubic centimeter of interstellar space, or ~1 eV/cm^3, which is comparable to the energy density of visible starlight at 0.3 eV/cm^3, the galactic magnetic field energy density (assumed 3 microgauss) which is ~0.25 eV/cm^3, or the cosmic microwave background (CMB) radiation energy density at ~ 0.25 eV/cm^3.[57]

5.4   Detection methods

The VERITAS array of air Cherenkov telescopes.

There are several ground-based methods of detecting cosmic rays currently in use. The first detection method is called the air Cherenkov telescope, designed to detect low-energy (<200 GeV) cosmic rays by means of analyzing their Cherenkov radiation, which for cosmic rays are gamma rays emitted as they travel faster than the speed of light in their medium, the atmosphere.[58] While these telescopes are extremely good at distinguishing between background radiation and that of cosmic-ray origin, they can only function well on clear nights without the Moon shining, and have very small fields of view and are only active for a few percent of the time. Another Cherenkov telescope uses water as a medium through which particles pass and produce Cherenkov radiation to make them detectable.[59]

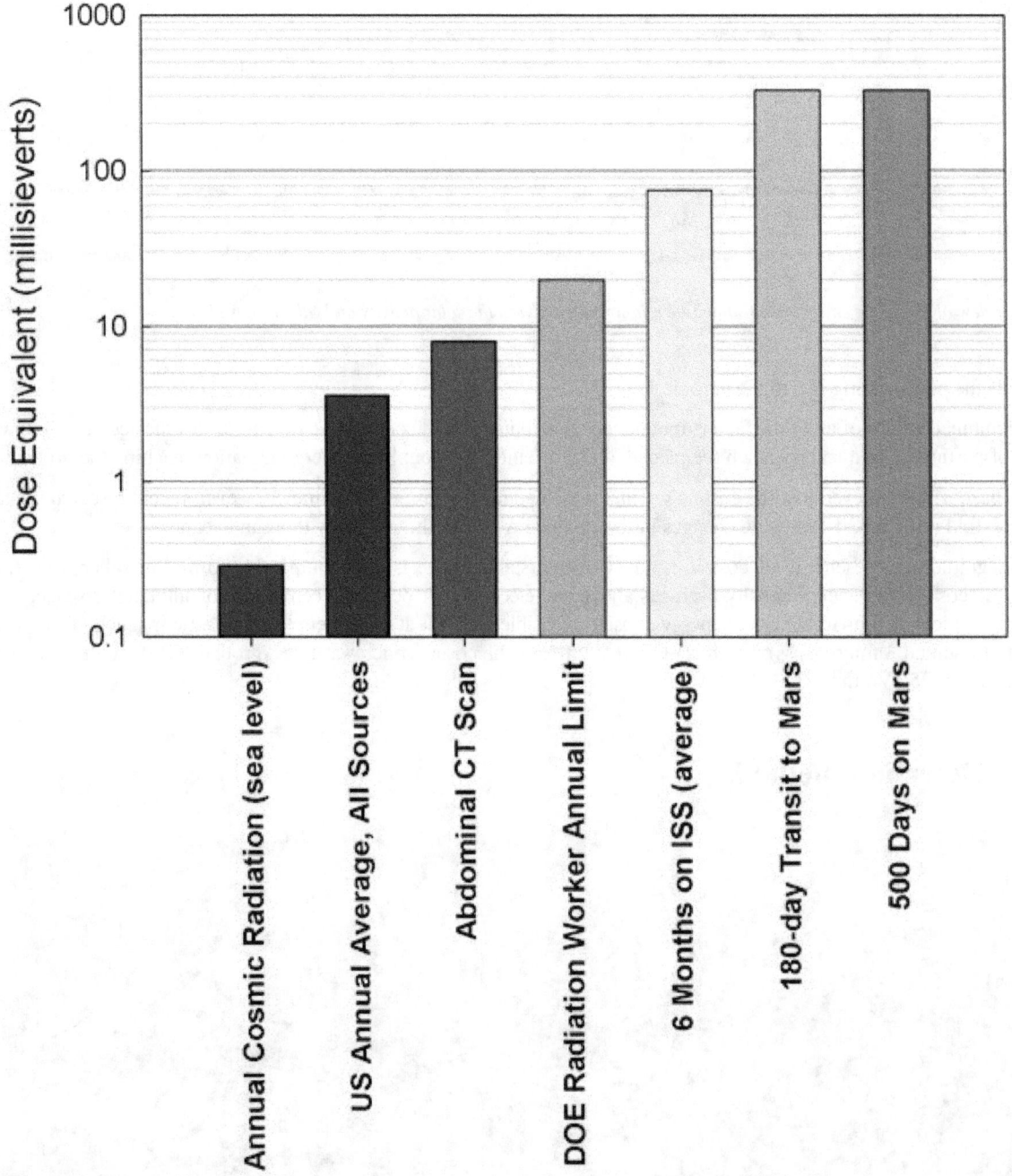

Comparison of radiation doses - includes the amount detected on the trip from Earth to Mars by the RAD on the MSL (2011 - 2013).[17][18][19]

Extensive air shower (EAS) arrays, a second detection method, measure the charged particles which pass through them.

EAS arrays measure much higher-energy cosmic rays than air Cherenkov telescopes, and can observe a broad area of the sky and can be active about 90% of the time. However, they are less able to segregate background effects from cosmic rays than can air Cherenkov telescopes. EAS arrays employ plastic scintillators in order to detect particles.

Another method was developed by Robert Fleischer, P. Buford Price, and Robert M. Walker for use in high-altitude balloons.[60] In this method, sheets of clear plastic, like 0.25 mm Lexan polycarbonate, are stacked together and exposed directly to cosmic rays in space or high altitude. The nuclear charge causes chemical bond breaking or ionization in the plastic. At the top of the plastic stack the ionization is less, due to the high cosmic ray speed. As the cosmic ray speed decreases due to deceleration in the stack, the ionization increases along the path. The resulting plastic sheets are "etched" or slowly dissolved in warm caustic sodium hydroxide solution, that removes the surface material at a slow, known rate. The caustic sodium hydroxide dissolves the plastic at a faster rate along the path of the ionized plastic. The net result is a conical etch pit in the plastic. The etch pits are measured under a high-power microscope (typically 1600x oil-immersion), and the etch rate is plotted as a function of the depth in the stacked plastic.

This technique yields a unique curve for each atomic nucleus from 1 to 92, allowing identification of both the charge and energy of the cosmic ray that traverses the plastic stack. The more extensive the ionization along the path, the higher the charge. In addition to its uses for cosmic-ray detection, the technique is also used to detect nuclei created as products of nuclear fission.

A fourth method involves the use of cloud chambers[61] to detect the secondary muons created when a pion decays. Cloud chambers in particular can be built from widely available materials and can be constructed even in a high-school laboratory. A fifth method, involving bubble chambers, can be used to detect cosmic ray particles.[62]

Another method detects the light from nitrogen fluorescence caused by the excitation of nitrogen in the atmosphere by the shower of particles moving through the atmosphere. This method allows for accurate detection of the direction from which the cosmic ray came.[63]

Finally, the CMOS devices in pervasive smartphone cameras have been proposed as a practical distributed network to detect air showers from ultra-high energy cosmic rays (UHECRs) which is at least comparable with that of conventional cosmic ray detectors.[64] The app, which is currently in beta and accepting applications, is CRAYFIS (Cosmic RAYs Found In Smartphones).[65][66]

5.5   Effects

5.5.1   Changes in atmospheric chemistry

Cosmic rays ionize the nitrogen and oxygen molecules in the atmosphere, which leads to a number of chemical reactions. One of the reactions results in ozone depletion. Cosmic rays are also responsible for the continuous production of a number of unstable isotopes in the Earth's atmosphere, such as carbon-14, via the reaction:

$$n + {}^{14}N \rightarrow p + {}^{14}C$$

Cosmic rays kept the level of carbon-14[67] in the atmosphere roughly constant (70 tons) for at least the past 100,000 years, until the beginning of above-ground nuclear weapons testing in the early 1950s. This is an important fact used in radiocarbon dating used in archaeology.

Reaction products of primary cosmic rays, radioisotope half-lifetime, and production reaction.[68]

- Tritium (12.3 years): 14N(n, 3H)12C (Spallation)

- Beryllium-7 (53.3 days)

- Beryllium-10 (1.39 million years): 14N(n,p α)10Be (Spallation)

- Carbon-14 (5730 years): 14N(n, p)14C (Neutron activation)

- Sodium-22 (2.6 years)

- Sodium-24 (15 hours)

- Magnesium-28 (20.9 hours)

- Silicon-31 (2.6 hours)

- Silicon-32 (101 years)

- Phosphorus-32 (14.3 days)

- Sulfur-35 (87.5 days)

- Sulfur-38 (2.84 hours)

- Chlorine-34 m (32 minutes)

- Chlorine-36 (300,000 years)

- Chlorine-38 (37.2 minutes)

- Chlorine-39 (56 minutes)

- Argon-39 (269 years)

- Krypton-85 (10.7 years)

5.5.2 Role in ambient radiation

Cosmic rays constitute a fraction of the annual radiation exposure of human beings on the Earth, averaging 0.39 mSv out of a total of 3 mSv per year (13% of total background) for the Earth's population. However, the background radiation from cosmic rays increases with altitude, from 0.3 mSv per year for sea-level areas to 1.0 mSv per year for higher-altitude cities, raising cosmic radiation exposure to a quarter of total background radiation exposure for populations of said cities. Airline crews flying long distance high-altitude routes can be exposed to 2.2 mSv of extra radiation each year due to cosmic rays, nearly doubling their total ionizing radiation exposure.

> Figures are for the time before the Fukushima Daiichi nuclear disaster. Human-made values by UNSCEAR are from the Japanese National Institute of Radiological Sciences, which summarized the UNSCEAR data.

5.5.3 Effect on electronics

Cosmic rays have sufficient energy to alter the states of circuit components in electronic integrated circuits, causing transient errors to occur, such as corrupted data in electronic memory devices, or incorrect performance of CPUs, often referred to as "soft errors" (not to be confused with software errors caused by programming mistakes/bugs). This has been a problem in electronics at extremely high-altitude, such as in satellites, but with transistors becoming smaller and smaller, this is becoming an increasing concern in ground-level electronics as well.[74] Studies by IBM in the 1990s suggest that computers typically experience about one cosmic-ray-induced error per 256 megabytes of RAM per month.[75] To alleviate this problem, the Intel Corporation has proposed a cosmic ray detector that could be integrated into future high-density microprocessors, allowing the processor to repeat the last command following a cosmic-ray event.[76]

Cosmic rays are suspected as a possible cause of an in-flight incident in 2008 where an Airbus A330 airliner of Qantas twice plunged hundreds of feet after an unexplained malfunction in its flight control system. Many passengers and crew members were injured, some seriously. After this incident, the accident investigators determined that the airliner's flight control system had received a data spike that could not be explained, and that all systems were in perfect working order. This has prompted a software upgrade to all A330 and A340 airliners, worldwide, so that any data spikes in this system are filtered out electronically.[77]

See also: radiation hardening
See also: ECC memory

5.5.4   Significance to space travel

Main article: Health threat from cosmic rays

Galactic cosmic rays are one of the most important barriers standing in the way of plans for interplanetary travel by crewed spacecraft. Cosmic rays also pose a threat to electronics placed aboard outgoing probes. In 2010, a malfunction aboard the Voyager 2 space probe was credited to a single flipped bit, probably caused by a cosmic ray. Strategies such as physical or magnetic shielding for spacecraft have been considered in order to minimize the damage to electronics and human beings caused by cosmic rays.[78][79]

5.5.5   Role in lightning

Cosmic rays have been implicated in the triggering of electrical breakdown in lightning. It has been proposed that essentially all lightning is triggered through a relativistic process, "runaway breakdown", seeded by cosmic ray secondaries. Subsequent development of the lightning discharge then occurs through "conventional breakdown" mechanisms.[80]

5.5.6   Postulated role in climate change

A role of cosmic rays directly or via solar-induced modulations in climate change was suggested by Edward P. Ney in 1959[81] and by Robert E. Dickinson in 1975.[82] The idea has been revived in recent years, most notably by Henrik Svensmark, who has argued that because solar variation modulates the cosmic ray flux on Earth, they would consequently affect the rate of cloud formation and hence the climate.[83][84] However, other scientists have vigorously criticized Svensmark for sloppy and inconsistent work: one example is adjustment of cloud data that understates error in lower cloud data, but not in high cloud data;[85] another example is "incorrect handling of the physical data" resulting in graphs that do not show the correlations they claim to show.[86]

In any case, 97% of climate scientists[87] support the conclusions in the 2007 IPCC synthesis report, which strongly attributes a major role in the ongoing global warming to human-produced gases such as carbon dioxide, methane, nitrous oxide, and halocarbons, and has stated that models including natural forcings only (including aerosol forcings, which cosmic rays are considered by some to contribute to) would result in far less warming than has actually been observed or predicted in models including anthropogenic forcings.[88]

Svensmark is one of several scientists outspokenly opposed to the mainstream scientific assessment of global warming. His estimates on the magnitude of the effect of GCR (galactic cosmic rays) on global warming continue to be refuted in the mainstream scientific press.[89] For instance, a November 2013 study showed that less than 14 percent of global warming since the 1950s could be attributed to cosmic ray rate, and while the models showed a small correlation every 22 years, the cosmic ray rate did not match the changes in temperature, indicating that it was not a causal relationship.[90] Another 2013 study found, contrary to Svensmark's claims, "no statistically significant correlations between cosmic rays and global albedo or globally averaged cloud height."[91]

5.6   Research and experiments

See also: Cosmic-ray observatory

There are a number of cosmic-ray research initiatives.

5.6.1 Ground-based

- Akeno Giant Air Shower Array
- CHICOS
- CRIPT
- High Energy Stereoscopic System
- High Resolution Fly's Eye Cosmic Ray Detector
- MAGIC
- MARIACHI
- Pierre Auger Observatory
- Telescope Array Project
- Washington Large Area Time Coincidence Array
- CLOUD
- Spaceship Earth
- Milagro
- NMDB
- KASCADE
- GAMMA
- GRAPES-3
- HEGRA
- Chicago Air Shower Array
- IceCube

5.6.2 Satellite

- PAMELA
- Alpha Magnetic Spectrometer
- ACE (Advanced Composition Explorer)
- Voyager 1 and Voyager 2
- Cassini–Huygens
- HEAO 1, HEAO 2, HEAO 3
- Fermi Gamma-ray Space Telescope
- Solar and Heliospheric Observatory
- Interstellar Boundary Explorer
- Langton Ultimate Cosmic-Ray Intensity Detector

5.6.3   Balloon-borne

- BESS

- Advanced Thin Ionization Calorimeter

- TRACER (cosmic ray detector)

- TIGER

- Cosmic Ray Energetics and Mass (CREAM)

- PERDaix

- HEAT (High Energy Antimatter Telescope)

5.7   See also

- Environmental radioactivity

- Forbush decrease

- Gilbert Jerome Perlow

- Extragalactic cosmic ray

- Solar energetic particle

- Track Imaging Cherenkov Experiment

- Cosmic ray visual phenomena

- Health threat from cosmic rays

- Central nervous system effects from radiation exposure during spaceflight

5.8   References

[1] Sharma (2008). *Atomic And Nuclear Physics*. Pearson Education India. p. 478. ISBN 978-81-317-1924-4.

[2] Ackermann, M.; Ajello, M.; Allafort, A.; Baldini, L.; Ballet, J.; Barbiellini, G.; Baring, M. G.; Bastieri, D.; Bechtol, K.; Bellazzini, R.; Blandford, R. D.; Bloom, E. D.; Bonamente, E.; Borgland, A. W.; Bottacini, E.; Brandt, T. J.; Bregeon, J.; Brigida, M.; Bruel, P.; Buehler, R.; Busetto, G.; Buson, S.; Caliandro, G. A.; Cameron, R. A.; Caraveo, P. A.; Casandjian, J. M.; Cecchi, C.; Celik, O.; Charles, E.; et al. (2013-02-15). "Detection of the Characteristic Pion-Decay Signature in Supernova Remnants". *Science* (American Association for the Advancement of Science) **339** (6424): 807–811. arXiv:1302.3307. Bibcode:2013Sci...339..807A. doi:10.1126/science.1231160. Retrieved 2013-02-14.

[3] Ginger Pinholster (2013-02-13). "Evidence Shows that Cosmic Rays Come from Exploding Stars".

[4] Dr. Eric Christian. "Are Cosmic Rays Electromagnetic radiation?". NASA. Retrieved 2012-12-11.

[5] Nerlich, Steve (12 June 2011). "Astronomy Without A Telescope – Oh-My-God Particles". *Universe Today*. Universe Today. Retrieved 17 February 2013.

[6] "Facts and figures". *The LHC*. European Organization for Nuclear Research. 2008. Retrieved 17 February 2013.

[7] Gaensler, Brian (November 2011). "Extreme speed". *COSMOS* (41).

[8] L. Anchordoqui, T. Paul, S. Reucroft, J. Swain; Paul; Reucroft; Swain (2003). "Ultrahigh Energy Cosmic Rays: The state of the art before the Auger Observatory". *International Journal of Modern Physics A* **18** (13): 2229–2366. arXiv:hep-ph/0206072. Bibcode:2003IJMPA..18.2229A. doi:10.1142/S0217751X03013879.

[9] Nave, Carl R. "Cosmic rays". *HyperPhysics Concepts.* Georgia State University. Retrieved 17 February 2013.

[10] "What are cosmic rays?". NASA, Goddard Space Flight Center. Retrieved 31 October 2012. copy

[11] Malley, Marjorie C. (August 25, 2011), *Radioactivity: A History of a Mysterious Science*, Oxford University Press, pp. 78–79.

[12] North, John (July 15, 2008), *Cosmos: An Illustrated History of Astronomy and Cosmology*, University of Chicago Press, p. 686.

[13] D. Pacini (1912). "La radiazione penetrante alla superficie ed in seno alle acque". *Il Nuovo Cimento, Series VI* **3**: 93–100. doi:10.1007/BF02957440.

> Translated and commented in A. de Angelis (2010). "Penetrating Radiation at the Surface of and in Water". *Nuovo Cimento VI/ 9 (1912)* **3** (93). arXiv:1002.1810. Bibcode:2010arXiv1002.1810P.

[14] "Nobel Prize in Physics 1936 – Presentation Speech". Nobelprize.org. 1936-12-10. Retrieved 2013-02-27.

[15] V.F. Hess (1936). "The Nobel Prize in Physics 1936". The Nobel Foundation. Retrieved 2010-02-11.

[16] V.F. Hess (1936). "Unsolved Problems in Physics: Tasks for the Immediate Future in Cosmic Ray Studies". *Nobel Lectures.* The Nobel Foundation. Retrieved 2010-02-11.

[17] Kerr, Richard (31 May 2013). "Radiation Will Make Astronauts' Trip to Mars Even Riskier". *Science* **340** (6136): 1031. doi:10.1126/science.340.6136.1031. Retrieved 31 May 2013.

[18] Zeitlin, C.; Hassler, D. M.; Cucinotta, F. A.; Ehresmann, B.; Wimmer-Schweingruber, R. F.; Brinza, D. E.; Kang, S.; Weigle, G.; et al. (31 May 2013). "Measurements of Energetic Particle Radiation in Transit to Mars on the Mars Science Laboratory". *Science* **340** (6136): 1080–1084. Bibcode:2013Sci...340.1080Z. doi:10.1126/science.1235989. Retrieved 31 May 2013.

[19] Chang, Kenneth (30 May 2013). "Data Point to Radiation Risk for Travelers to Mars". New York Times. Retrieved 31 May 2013.

[20] Clay, J. (1927). "Title unknown". *Proceedings of the Section of Sciences, Koninklijke Akademie van Wetenschappen te Amsterdam* **30**: 633.

[21] Bothe, Walther and Werner Kolhörster (November 1929). "Das Wesen der Höhenstrahlung". *Zeitschrift für Physik* **56** (11–12): 751–777. Bibcode:1929ZPhy...56..751B. doi:10.1007/BF01340137.

[22] Rossi, Bruno (August 1930)."On the Magnetic Deflection of Cosmic Rays".*Physical Review***36**(3): 606.Bibcode:1930PhRv... doi:10.1103/PhysRev.36.606.

[23] Johnson, Thomas H. (May 1933). "The Azimuthal Asymmetry of the Cosmic Radiation". *Physical Review* **43** (10): 834–835. Bibcode:1933PhRv...43..834J. doi:10.1103/PhysRev.43.834.

[24] Alvarez, Luis; Compton, Arthur Holly; Compton (May 1933). "A Positively Charged Component of Cosmic Rays". *Physical Review* **43** (10): 835–836. Bibcode:1933PhRv...43..835A. doi:10.1103/PhysRev.43.835.

[25] Rossi, Bruno (May 1934). "Directional Measurements on the Cosmic Rays Near the Geomagnetic Equator". *Physical Review* **45** (3): 212–214. Bibcode:1934PhRv...45..212R. doi:10.1103/PhysRev.45.212.

[26] Freier, Phyllis; Lofgren, E.; Ney, E.; Oppenheimer, F.; Bradt, H.; Peters, B.; et al. (July 1948). "Evidence for Heavy Nuclei in the Primary Cosmic radiation". *Physical Review* **74** (2): 213–217. Bibcode:1948PhRv...74..213F. doi:10.1103/PhysRev.74.213.

[27] Freier, Phyllis; Peters, B.; et al. (December 1948). "Investigation of the Primary Cosmic Radiation with Nuclear Photographic Emulsions". *Physical Review* **74** (12): 1828–1837. Bibcode:1948PhRv...74.1828B. doi:10.1103/PhysRev.74.1828.

[28] Auger, P.; et al. (July 1939), "Extensive Cosmic-Ray Showers",*Reviews of Modern Physics***11**(3–4): 288–291,Bibcode:1939R doi:10.1103/RevModPhys.11.288.

[29] J.L. DuBois, R.P. Multhauf, C.A. Ziegler (2002). *The Invention and Development of the Radiosonde* (PDF). Smithsonian Studies in History and Technology **53**. Smithsonian Institution Press.

[30] S. Vernoff (1935). "Radio-Transmission of Cosmic Ray Data from the Stratosphere". *Nature* **135** (3426): 1072–1073. Bibcode:1935Natur.135.1072V. doi:10.1038/1351072c0.

[31] Clark, G.; Earl, J.; Kraushaar, W.; Linsley, J.; Rossi, B.; Scherb, F.; Scott, D. (1961). "Cosmic-Ray Air Showers at Sea Level". *Physical Review* **122** (2): 637. Bibcode:1961PhRv..122..637C. doi:10.1103/PhysRev.122.637.

[32] Auger Project. "Auger Observatory: A New Astrophysics Facility Rises from the Pampa". *Pierre Auger Observatory*. Auger Project. Retrieved 2013-04-29.

[33] Kraushaar, W. L; et al. (1972). "Title unknown". *The Astrophysical Journal* **177**: 341. Bibcode:1972ApJ...177..341K. doi:10.1086/151713.

[34] Baade, W.; Zwicky, F. (1934). "Cosmic Rays from Super-novae". *Proceedings of the National Academy of Sciences of the United States of America* (National Academy of Sciences) **20** (5): 259–263. Bibcode:1934PNAS...20..259B. doi:10.1073/pnas.20.5.259. JSTOR 86841.

[35]Babcock, H. (1948). "Magnetic Variable Stars as Sources of Cosmic Rays".*Physical Review***74**(4): 489.Bibcode:1948PhRv...74 doi:10.1103/PhysRev.74.489.

[36] Sekido, Y.; Masuda, T.; Yoshida, S.; Wada, M. (1951). "The Crab Nebula as an Observed Point Source of Cosmic Rays". *Physical Review* **83** (3): 658. Bibcode:1951PhRv...83..658S. doi:10.1103/PhysRev.83.658.2.

[37] Gibb, Meredith (3 February 2010). "Cosmic Rays". *Imagine the Universe*. NASA Goddard Space Flight Center. Retrieved 17 March 2013.

[38] Hague, J. D. (July 2009). "Correlation of the Highest Energy Cosmic Rays with Nearby Extragalactic Objects in Pierre Auger Observatory Data" (PDF). *Proceedings of the 31st ICRC, Łódź 2009*. International Cosmic Ray Conference. Łódź, Poland. pp. 6–9. Retrieved 17 March 2013.

[39] Hague, J. D. (July 2009). "Correlation of the Highest Energy Cosmic Rays with Nearby Extragalactic Objects in Pierre Auger Observatory Data" (PDF). *Proceedings of the 31st ICRC, Łódź, Poland 2009 - International Cosmic Ray Conference*: 36–39. Retrieved 17 March 2013.

[40] Moskowitz, Clara (25 June 2009). "Source of Cosmic Rays Pinned Down". *Space.com*. TechMediaNetwork. Retrieved 20 March 2013.

[41] Adriani, O.; Barbarino, G. C.; Bazilevskaya, G. A.; Bellotti, R.; Boezio, M.; Bogomolov, E. A.; Bonechi, L.; Bongi, M.; Bonvicini, V.; Borisov, S.; Bottai, S.; Bruno, A.; Cafagna, F.; Campana, D.; Carbone, R.; Carlson, P.; Casolino, M.; Castellini, G.; Consiglio, L.; De Pascale, M. P.; De Santis, C.; De Simone, N.; Di Felice, V.; Galper, A. M.; Gillard, W.; Grishantseva, L.; Jerse, G.; Karelin, A. V.; Koldashov, S. V.; Krutkov, S. Y. (2011). "PAMELA Measurements of Cosmic-Ray Proton and Helium Spectra". *Science* **332** (6025): 69–72. arXiv:1103.4055. Bibcode:2011Sci...332...69A. doi:10.1126/science.1199172. PMID 21385721.

[42] Jha, Alok (14 February 2013). "Cosmic ray mystery solved". *The Guardian*. Guardian News and Media Limited. Retrieved 21 March 2013.

[43] Mewaldt, R.A., 2010. "Cosmic Rays". California Institute of Technology.

[44] Koch, L.; Engelmann, J. J.; Goret, P.; Juliusson, E.; Petrou, N.; Rio, Y.; Soutoul, A.; Byrnak, B.; Lund, N.; Peters, B.; Engelmann; Goret; Juliusson; Petrou; Rio; Soutoul; Byrnak; Lund; Peters (October 1981). "The relative abundances of the elements scandium to manganese in relativistic cosmic rays and the possible radioactive decay of manganese 54". *Astronomy and Astrophysics* **102** (11): L9. Bibcode:1981A&A...102L...9K.

[45] L. Accardo (AMS Collaboration); et al. (18 September 2014). "High Statistics Measurement of the Positron Fraction in Primary Cosmic Rays of 0.5–500 GeV with the Alpha Magnetic Spectrometer on the International Space Station" (PDF). *Physical Review Letters* **113**: 121101. Bibcode:2014PhRvL.113l1101A. doi:10.1103/PhysRevLett.113.121101.

[46] Schirber, Michael. "Synopsis: More Dark Matter Hints from Cosmic Rays?". American Physical Society. Retrieved 21 September 2014.

[47] "New results from the Alpha Magnetic$Spectrometer on the International Space Station" (PDF). *AMS-02 at NASA*. Retrieved 21 September 2014.

[48] Aguilar, M.; Alberti, G.; Alpat, B.; Alvino, A.; Ambrosi, G.; Andeen, K.; Anderhub, H.; Arruda, L.; Azzarello, P.; Bachlechner, A.; Barao, F.; Baret, B.; Barrau, A.; Barrin, L.; Bartoloni, A.; Basara, L.; Basili, A.; Batalha, L.; Bates, J.; Battiston, R.; Bazo, J.; Becker, R.; Becker, U.; Behlmann, M.; Beischer, B.; Berdugo, J.; Berges, P.; Bertucci, B.; Bigongiari, G.; et al. (2013). "First Result from the Alpha Magnetic Spectrometer on the International Space Station: Precision Measurement of the Positron Fraction in Primary Cosmic Rays of 0.5–350 GeV". *Physical Review Letters* **110** (14): 141102. Bibcode:2013PhRvL.110n1102A. doi:10.1103/PhysRevLett.110.141102.

[49] Moskalenko, I. V.; Strong, A. W.; Ormes, J. F; Potgieter, M. S. (January 2002). "Secondary antiprotons and propagation of cosmic rays in the Galaxy and heliosphere". *The Astrophysical Journal* **565** (1): 280–296. arXiv:astro-ph/0106567v2. Bibcode:2002ApJ...565..280M. doi:10.1086/324402.

[50] AMS Collaboration; Aguilar, M.; Alcaraz, J.; Allaby, J.; Alpat, B.; Ambrosi, G.; Anderhub, H.; Ao, L.; et al. (August 2002). "The Alpha Magnetic Spectrometer (AMS) on the International Space Station: Part I – results from the test flight on the space shuttle". *Physics Reports* **366** (6): 331–405. Bibcode:2002PhR...366..331A. doi:10.1016/S0370-1573(02)00013-3.

[51] "EGRET Detection of Gamma Rays from the Moon". NASA/GSFC. 1 August 2005. Retrieved 2010-02-11.

[52] Morison, Ian (2008). *Introduction to Astronomy and Cosmology*. John Wiley & Sons. p. 198. ISBN 978-0-470-03333-3.

[53] "Extreme Space Weather Events". National Geophysical Data Center.

[54] "Pierre Auger Observatory". Auger.org. Retrieved 2012-08-17.

[55] "Pierre Auger Observatory". Auger.org. Retrieved 2015-07-15.

[56] D. Lal, A.J.T. Jull, D. Pollard, L. Vacher; Jull; Pollard; Vacher (2005). "Evidence for large century time-scale changes in solar activity in the past 32 Kyr, based on in-situ cosmogenic ^{14}C in ice at Summit, Greenland". *Earth and Planetary Science Letters* **234** (3–4): 335–249. Bibcode:2005E&PSL.234..335L. doi:10.1016/j.epsl.2005.02.011.

[57] Castellina, Antonella; Donato, Fiorenza (2012). Oswalt, T.D McLean, I.S.; Bond, H.E.; French, L.; Kalas, P.; Barstow, M.; Gilmore,G.F.; Keel, W., ed. *Planets, Stars, and Stellar Systems* (1 ed.). Springer. ISBN 978-90-481-8817-8.

[58] "The Detection of Cosmic Rays". *Milagro Gamma-Ray Observatory*. Los Alamos National Laboratory. 3 April 2002. Retrieved 22 February 2013.

[59] "What are cosmic rays?" (PDF). Michigan State University National Superconducting Cyclotron Laboratory. Retrieved 23 February 2013.

[60] R.L. Fleischer, P.B. Price, R.M. Walker (1975). *Nuclear tracks in solids: Principles and applications*. University of California Press.

[61] "Cloud Chambers and Cosmic Rays: A Lesson Plan and Laboratory Activity for the High School Science Classroom" (PDF). Cornell University Laboratory for Elementary-Particle Physics. 2006. Retrieved 23 February 2013.

[62] Chu, W.; Kim, Y.; Beam, W.; Kwak, N. (1970). "Evidence of a Quark in a High-Energy Cosmic-Ray Bubble-Chamber Picture". *Physical Review Letters* **24** (16): 917. Bibcode:1970PhRvL..24..917C. doi:10.1103/PhysRevLett.24.917.

[63] Letessier-Selvon, Antoine; Stanev, Todor. "Ultrahigh energy cosmic rays". *Reviews of Modern Physics* **83** (3): 907–942. arXiv:1103.0031. Bibcode:2011RvMP...83..907L. doi:10.1103/RevModPhys.83.907.

[64] /cosmic-ray-particle-shower-theres-an-app-for-that/

[65] Collaboration website

[66] CRAYFIS detector array paper.

[67] Trumbore, Susan (2000). Noller, J. S., J. M. Sowers, and W. R. Lettis, ed. *Quaternary Geochronology: Methods and Applications*. Washington, D.C.: American Geophysical Union. pp. 41–59. ISBN 0-87590-950-7.

[68] "Natürliche, durch kosmische Strahlung laufend erzeugte Radionuklide" (PDF) (in German). Retrieved 2010-02-11.

[69] UNSCEAR "Sources and Effects of Ionizing Radiation" page 339 retrieved 2011-6-29

[70] Japan NIRS UNSCEAR 2008 report page 8 retrieved 2011-6-29

[71] Princeton.edu "Background radiation" retrieved 2011-6-29

[72] Washington state Dept. of Health "Background radiation" retrieved 2011-6-29

[73] Ministry of Education, Culture, Sports, Science, and Technology of Japan "Radiation in environment" retrieved 2011-6-29

[74] *IBM experiments in soft fails in computer electronics (1978–1994)*, from *Terrestrial cosmic rays and soft errors*, IBM Journal of Research and Development, Vol. 40, No. 1, 1996. Retrieved 16 April 2008.

[75] Scientific American (2008-07-21). "Solar Storms: Fast Facts". Nature Publishing Group. Retrieved 2009-12-08.

[76] *Intel plans to tackle cosmic ray threat*, BBC News Online, 8 April 2008. Retrieved 16 April 2008.

[77] *Cosmic rays may have hit Qantas plane off the coast of North West Australia*, News.com.au, 18 November 2009. Retrieved 19 November 2009.

[78] Globus, Al (10 July 2002). "Appendix E: Mass Shielding". *Space Settlements: A Design Study*. NASA. Retrieved 24 February 2013.

[79] Atkinson, Nancy (24 January 2005). "Magnetic shielding for spacecraft". *The Space Review*. Retrieved 24 February 2013.

[80] *Runaway Breakdown and the Mysteries of Lightning*, Physics Today, May 2005.

[81] Ney, Edward P. (14 February 1959)."Cosmic Radiation and the Weather".*Nature***183**(4659): 451–452.Bibcode:1959Natur.183 doi:10.1038/183451a0. Retrieved 2012-02-09.

[82] Dickinson, Robert E. (December 1975). "Solar Variability and the Lower Atmosphere". *Bulletin of the American Meteorological Society* **56** (12): 1240–1248. Bibcode:1975BAMS...56.1240D. doi:10.1175/1520-0477(1975)056<1240:SVATLA>2.0.CO;2.

[83] Long, Marion (25 June 2007). "Sun's Shifts May Cause Global Warming". Discover. Retrieved 7 July 2013.

[84] Henrik Svensmark (1998). "Influence of Cosmic Rays on Earth's Climate". *Physical Review Letters* **81** (22): 5027–5030. Bibcode:1998PhRvL..81.5027S. doi:10.1103/PhysRevLett.81.5027.

[85] Benestad, Rasmus E. "'Cosmoclimatology' – tired old arguments in new clothes". Retrieved 13 November 2013.

[86] Peter Laut, "Solar activity and terrestrial climate: an analysis of some purported correlations", Journal of Atmospheric and Solar-Terrestrial Physics 65 (2003) 801- 812

[87] Anderegg, William R. L.; Prall, J. W.; Harold, J.; Schneider, S. H. (21 June 2010). "Expert credibility in climate change". *Proceedings of the National Academy of Sciences of the United States of America* **107** (27): 12107–12109. Bibcode:2010PNAS..10 712107A.doi:10.1073/pnas.1003187107.PMC2901439.PMID20566872.Retrieved7July2013.

[88] Core Writing Team (17 November 2007). Pachauri, R.K. and Reisinger, A., ed. *Climate Change 2007: Synthesis Report*. IPCC Plenary XXVII. Valencia, Spain: Intergovernmental Panel on Climate Change. pp. 39–40. Retrieved 24 February 2013.

[89] Plait, Phil (31 August 2011). "No, a new study does not show cosmic-rays are connected to global warming". *Discover*. Retrieved 7 July 2013.

[90] Sloan, T.; Wolfendale, A.W. (November 7, 2013). "Cosmic rays, solar activity and the climate". *Environmental Research Letters* **8**: 045022. Bibcode:2013ERL.....8d5022S. doi:10.1088/1748-9326/8/4/045022.

[91] Krissansen-Totton, J.; Davies, R. (2013). "Investigation of Cosmic Ray-Cloud Connections Using MISR". *Geophysical Research Letters* **40**: 5240–5245. arXiv:1311.1308. Bibcode:2013GeoRL..40.5240K. doi:10.1002/grl.50996.

5.9   Further references

- R.G. Harrison and D.B. Stephenson, Detection of a galactic cosmic ray influence on clouds, Geophysical Research Abstracts, Vol. 8, 07661, 2006 SRef-ID: 1607-7962/gra/EGU06-A-07661

- Anderson, C. D.; Neddermeyer, S. H. (1936). "Cloud Chamber Observations of Cosmic Rays at 4300 Meters Elevation and Near Sea-Level". *Phys. Rev* **50**: 263–271. Bibcode:1936PhRv...50..263A. doi:10.1103/physrev.50.263.

- Boezio, M.; et al. (2000). "Measurement of the flux of atmospheric muons with the CAPRICE94 apparatus". *Phys. Rev. D* **62**: 032007. arXiv:hep-ex/0004014. Bibcode:2000PhRvD..62c2007B. doi:10.1103/physrevd.62.032007.

- R. Clay and B. Dawson, Cosmic Bullets, Allen & Unwin, 1997. ISBN 1-86448-204-4

- T. K. Gaisser, *Cosmic Rays and Particle Physics*, Cambridge University Press, 1990. ISBN 0-521-32667-2

- P. K. F. Grieder, Cosmic Rays at Earth: Researcher's Reference Manual and Data Book, Elsevier, 2001. ISBN 0-444-50710-8

- A. M. Hillas, *Cosmic Rays*, Pergamon Press, Oxford, 1972 ISBN 0-08-016724-1

- Kremer, J.; et al. (1999). "Measurement of Ground-Level Muons at Two Geomagnetic Locations". *Phys. Rev. Lett.* **83**: 4241–4244. Bibcode:1999PhRvL..83.4241K. doi:10.1103/physrevlett.83.4241.

- Neddermeyer, S. H.; Anderson, C. D. (1937). "Note on the Nature of Cosmic-Ray Particles". *Phys. Rev.* **51**: 884–886. Bibcode:1937PhRv...51..884N. doi:10.1103/physrev.51.884.

- M. D. Ngobeni and M. S. Potgieter, Cosmic ray anisotropies in the outer heliosphere, Advances in Space Research, 2007.

- M. D. Ngobeni, Aspects of the modulation of cosmic rays in the outer heliosphere, M.Sc Dissertation, Northwest University (Potchefstroom campus) South Africa 2006.

- D. Perkins, Particle Astrophysics, Oxford University Press, 2003. ISBN 0-19-850951-0

- C. E. Rolfs and S. R. William, Cauldrons in the Cosmos, The University of Chicago Press, 1988. ISBN 0-226-72456-5

- B. B. Rossi, *Cosmic Rays*, McGraw-Hill, New York, 1964.

- Martin Walt, Introduction to Geomagnetically Trapped Radiation, 1994. ISBN 0-521-43143-3

- Taylor, M.; Molla, M. (2010). "Towards a unified source-propagation model of cosmic rays". *Pub. Astron. Soc. Pac.* **424**: 98.

- Ziegler, J. F. (1981). "The Background In Detectors Caused By Sea Level Cosmic Rays". *Nuclear Instruments and Methods* **191**: 419–424. Bibcode:1981NIMPR.191..419Z. doi:10.1016/0029-554x(81)91039-9.

- TRACER Long Duration Balloon Project: the largest cosmic ray detector launched on balloons.

- Carlson, Per; De Angelis, Alessandro (2011). "Nationalism and internationalism in science: the case of the discovery of cosmic rays". *European Physical Journal H* **35** (4): 309–329. arXiv:1012.5068. Bibcode:2010EPJH...35..309C.doi:10.1140/epjh/e2011-10033-6.

5.10 External links

- Aspera European network portal

- Animation about cosmic rays on astroparticle.org

- Helmholtz Alliance for Astroparticle Physics

- Particle Data Group review of Cosmic Rays by C. Amsler et al., Physics Letters B667, 1 (2008).

- Introduction to Cosmic Ray Showers by Konrad Bernlöhr.

- BBC news, Cosmic rays find uranium, 2003.

- BBC news, Rays to nab nuclear smugglers, 2005.

- BBC news, Physicists probe ancient pyramid (using cosmic rays), 2004.

- Shielding Space Travelers by Eugene Parker.

- Anomalous cosmic ray hydrogen spectra from Voyager 1 and 2

- Anomalous Cosmic Rays (From NASA's Cosmicopia)

- Review of Cosmic Rays

- "Who's Afraid of a Solar Flare? Solar activity can be surprisingly good for astronauts." 7 October 2005, at Science@NASA

- video of Muon detector in use at Smithsonian Air and Space Museum

- Dr. Lothar Frey "Cosmic rays and electronic devices" (**YouTube Video**) SpaceUp Stuttgart 2012

- ARMAS, Real-time cosmic-ray radiation measurements at aviation altitudes.

- Padilla, Antonio (Tony). "Where do Cosmic Rays come from?". *Sixty Symbols*. Brady Haran for the University of Nottingham.

Chapter 6

Positronium

Positronium (Ps) is a system consisting of an electron and its anti-particle, a positron, bound together into an *exotic atom*, specifically an *onium*. The system is unstable: the two particles annihilate each other to predominantly produce two or three gamma-rays, depending on the relative spin states. The orbit and energy levels of the two particles are similar to that of the hydrogen atom (which is a bound state of a proton and an electron). However, because of the reduced mass, the frequencies of the spectral lines are less than half of the corresponding hydrogen lines.

6.1 States

The ground state of positronium, like that of hydrogen, has two possible configurations depending on the relative orientations of the spins of the electron and the positron.

The *singlet* state, 1S
0, with antiparallel spins ($S = 0$, $Ms = 0$) is known as *para-positronium* (p-Ps). It has a mean lifetime of 125 picoseconds and decays preferentially into two gamma rays with energy of 511 keV each (in the center-of-mass frame). Detection of these photons allows to reconstruct the vertex of the decay and is used in the positron-emission tomography. Para-positronium can decay into any even number of photons (2, 4, 6, ...), but the probability quickly decreases with the number: the branching ratio for decay into 4 photons is $1.439(2) \times 10^{-6}$.[1]

Para-positronium lifetime in vacuum is approximately[1]

$$t_0 = \frac{2\hbar}{m_e c^2 \alpha^5} = 1.244 \times 10^{-10} \text{ s.}$$

The *triplet* state, $^3\text{S}_1$, with parallel spins ($S = 1$, $Ms = -1, 0, 1$) is known as *ortho-positronium* (o-Ps). It has a mean lifetime of 142.05 ± 0.02 ns,[2] and the leading decay is three gammas. Other modes of decay are negligible; for instance, the five-photons mode has branching ratio of ~1.0×10^{-6}.[3]

Ortho-positronium lifetime in vacuum can be calculated approximately as:[1]

$$t_1 = \frac{\frac{1}{2}9h}{2m_e c^2 \alpha^6 (\pi^2 - 9)} = 1.386 \times 10^{-7} \text{ s.}$$

However more accurate calculations with corrections to order $O(\alpha^2)$ yield a value of 7.04 μs^{-1} for the decay rate, corresponding to a lifetime of 1.42×10^{-7} s.[4][5]

Positronium in the 2S state is metastable having a lifetime of 1.1 μs against annihilation. The positronium created in such an excited state will quickly cascade down to the ground state, where annihilation will occur more quickly.

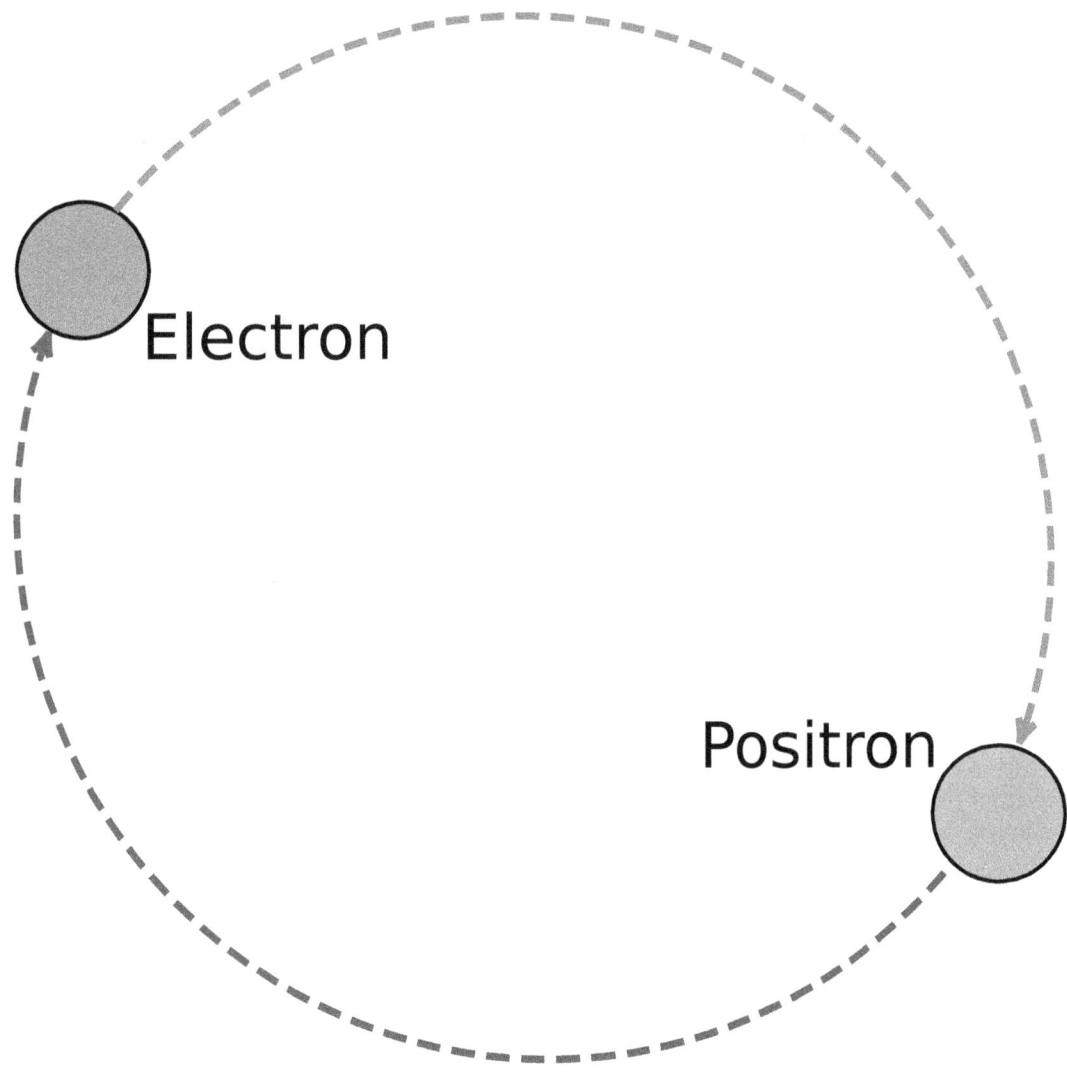

*An electron and positron orbiting around their common centre of mass. This is a bound quantum state known as **positronium**.*

6.1.1 Measurements

Measurements of these lifetimes and energy levels have been used in precision tests of quantum electrodynamics, confirming QED predictions to high precision.[1][6][7] Annihilation can proceed via a number of channels, each producing gamma rays with total energy of 1022 keV (sum of the electron and positron mass-energy), usually 2 or 3, with up to 5 recorded.

The annihilation into a neutrino–antineutrino pair is also possible, but the probability is predicted to be negligible. The branching ratio for o-Ps decay for this channel is 6.2×10^{-18} (electron neutrino–antineutrino pair) and 9.5×10^{-21} (for other flavour)[3] in predictions based on the Standard Model, but it can be increased by non-standard neutrino properties, like relatively high magnetic moment. The experimental upper limits on branching ratio for this decay (as well as for a decay into any "invisible" particles) are $<4.3 \times 10^{-7}$ for p-Ps and $<4.2 \times 10^{-7}$ for o-Ps.[2]

6.2 Energy levels

Main article: Bohr model § Electron energy levels

While precise calculation of positronium energy levels uses the Bethe–Salpeter equation or the Breit equation, the similarity between positronium and hydrogen allows a rough estimate. In this approximation, the energy levels are different because of a different effective mass, m^*, in the energy equation (see electron energy levels for a derivation):

$$E_n = -\frac{\mu q_e^4}{8h^2\varepsilon_0^2}\frac{1}{n^2},$$

where:

q_e is the charge magnitude of the electron (same as the positron),

h is Planck's constant,

ε_0 is the electric constant (otherwise known as the permittivity of free space),

μ is the reduced mass:

$$\mu = \frac{m_e m_p}{m_e + m_p} = \frac{m_e^2}{2m_e} = \frac{m_e}{2},$$

where m_e and m_p are, respectively, the mass of the electron and the positron (which are *the same* by definition as antiparticles).

Thus, for positronium, its reduced mass only differs from the electron by a factor of 2. This causes the energy levels to also roughly be half of what they are for the hydrogen atom.

So finally, the energy levels of positronium are given by

$$E_n = -\frac{1}{2}\frac{m_e q_e^4}{8h^2\varepsilon_0^2}\frac{1}{n^2} = \frac{-6.8\text{ eV}}{n^2}.$$

The lowest energy level of positronium ($n = 1$) is −6.8 electronvolts (eV). The next level is −1.7 eV. The negative sign implies a bound state. Positronium can also be considered by a particular form of the two-body Dirac equation; Two point particles with a Coulomb interaction can be exactly separated in the (relativistic) center-of-momentum frame and the resulting ground-state energy has been obtained very accurately using finite element methods of J. Shertzer.[8] The Dirac equation whose Hamiltonian comprises two Dirac particles and a static Coulomb potential is not relativistically invariant. But if one adds the $1/c^{2n}$ (or α^{2n}, where α is the fine-structure constant) terms, where $n = 1,2...$, then the result is relativistically invariant. Only the leading term is included. The α^2 contribution is the Breit term; workers rarely go to α^4 because at α^3 one has the Lamb shift, which requires quantum electrodynamics.[8]

6.3 History

Croatian scientist Stjepan Mohorovičić predicted the existence of positronium in a 1934 article published in *Astronomische Nachrichten*, in which he called it the "electrum".[9] Other sources credit Carl Anderson as having predicted its existence in 1932 while at Caltech.[10] It was experimentally discovered by Martin Deutsch at MIT in 1951 and became known as

The Positronium Beam at University College London, a lab used to study the properties of positronium

positronium.[10] Many subsequent experiments have precisely measured its properties and verified predictions of quantum electrodynamics (QED). There was a discrepancy known as the ortho-positronium lifetime puzzle that persisted for some time, but was eventually resolved with further calculations and measurements.[11] Measurements were in error because of the lifetime measurement of unthermalised positronium, which was only produced at a small rate. This had yielded lifetimes that were too long. Also calculations using relativistic QED are difficult to perform, so they had been done to only the first order. Corrections that involved higher orders were then calculated in a non-relativistic QED.[4]

6.4 Exotic compounds

Molecular bonding was predicted for positronium.[12] Molecules of positronium hydride (PsH) can be made.[13] Positronium can also form a cyanide and can form bonds with halogens or lithium.[14]

The first observation of di-positronium molecules—molecules consisting of two positronium atoms—was reported on 12 September 2007 by David Cassidy and Allen Mills from University of California, Riverside.[15][16]

6.5 Natural occurrence

Positronium in high energy states has been predicted to be the dominant form of atomic matter in the universe in the far future, if proton decay is a reality.[17]

6.6 See also

- Breit equation

- Antiprotonic helium

- Quantum electrodynamics

- Protonium

- Two-body Dirac equations

6.7 References

[1] Karshenboim, Savely G. (2003). "Precision Study of Positronium: Testing Bound State QED Theory". *International Journal of Modern Physics A [Particles and Fields; Gravitation; Cosmology; Nuclear Physics]* **19** (23): 3879–3896. arXiv:hep-ph/0310099. Bibcode:2004IJMPA..19.3879K. doi:10.1142/S0217751X04020142.

[2] A. Badertscher; et al. (2007). "An Improved Limit on Invisible Decays of Positronium". *Physical Review D* **75** (3): 032004. arXiv:hep-ex/0609059. Bibcode:2007PhRvD..75c2004B. doi:10.1103/PhysRevD.75.032004.

[3] Czarnecki, Andrzej; Karshenboim, Savely G. (2000). Levchenko, B.B.; Savrin, V.I., eds. "Decays of Positronium". *Proceedings of the International Workshop on High Energy Physics and Quantum Field Theory (QFTHEP)* (Moscow: MSU-Press) **14** (99): 538–544. arXiv:hep-ph/9911410. Bibcode:1999hep.ph...11410C.

[4] Kataoka, Y.; Asai, S.; Kobayashi, t. (9 September 2008). "First Test of $O(\alpha^2)$ Correction of the Orthopositronium Decay Rate" (PDF). International Center for Elementary Particle Physics.

[5] Adkins, G. S.; Fell, R. N.; Sapirstein, J. (29 May 2000). "Order α^2 Corrections to the Decay Rate of Orthopositronium". *Physical Review Letters* **84**(22): 5086–5089. arXiv:hep-ph/0003028. Bibcode:2000PhRvL..84.5086A. doi:10.1103/PhysRevLett.84

[6] Rubbia, A. (2004). "Positronium as a probe for new physics beyond the standard model". *International Journal of Modern Physics A [Particles and Fields; Gravitation; Cosmology; Nuclear Physics]* **19** (23): 3961–3985. arXiv:hep-ph/0402151. Bibcode:2004IJMPA..19.3961R. doi:10.1142/S0217751X0402021X.

[7] Vetter, P.A.; Freedman, S.J. (2002). "Branching-ratio measurements of multiphoton decays of positronium". *Physical Review A* **66** (5): 052505. Bibcode:2002PhRvA..66e2505V. doi:10.1103/PhysRevA.66.052505.

[8] Scott, T.C.; Shertzer, J.; Moore, R.A. (1992). "Accurate finite element solutions of the two-body Dirac equation". *Physical Review A* **45** (7): 4393–4398. Bibcode:1992PhRvA..45.4393S. doi:10.1103/PhysRevA.45.4393. PMID 9907514.

[9] Mohorovičić, S. (1934). "Möglichkeit neuer Elemente und ihre Bedeutung für die Astrophysik". *Astronomische Nachrichten* **253** (4): 94. Bibcode:1934AN....253...93M. doi:10.1002/asna.19342530402.

[10] "Martin Deutsch, MIT physicist who discovered positronium, dies at 85" (Press release). MIT. 2002.

[11] Dumé, Belle (May 23, 2003). "Positronium puzzle is solved". *Physics World*.

[12] Usukura, J.; Varga, K.; Suzuki, Y. (1998). "Signature of the existence of the positronium molecule". arXiv:physics/9804023v1 [physics.atom-ph].

[13] ""Out of This World" Chemical Compound Observed" (PDF). p. 9.

[14] Saito, Shiro L. (2000). "Is Positronium Hydride Atom or Molecule?". *Nuclear Instruments and Methods in Physics Research B* **171**: 60–66. Bibcode:2000NIMPB.171...60S. doi:10.1016/s0168-583x(00)00005-7.

[15] Cassidy, D.B.; Mills, A.P. (Jr.) (2007). "The production of molecular positronium". *Nature* **449**(7159): 195–197. Bibcode: doi:10.1038/nature06094. PMID 17851519. Lay summary.

[16] "Molecules of positronium observed in the lab for the first time". Physorg.com. Retrieved 2007-09-07.

[17] A dying universe: the long-term fate and evolution of astrophysical objects, Fred C. Adams and Gregory Laughlin, *Reviews of Modern Physics* **69**, #2 (April 1997), pp. 337–372. Bibcode: 1997RvMP...69..337A. doi:10.1103/RevModPhys.69.337 arXiv:astro-ph/9701131.

6.8   External links

- The Search for Positronium

- Obituary of Martin Deutsch, discoverer of Positronium

- Website about positrons, positronium and antihydrogen. Positron Laboratory, Como, Italy

Chapter 7

List of particles

This is a list of the different types of particles found or believed to exist in the whole of the universe. For individual lists of the different particles, see the list below.

7.1 Elementary particles

Main article: Elementary particle

Elementary particles are particles with no measurable internal structure; that is, they are not composed of other particles. They are the fundamental objects of quantum field theory. Many families and sub-families of elementary particles exist. Elementary particles are classified according to their spin. Fermions have half-integer spin while bosons have integer spin. All the particles of the Standard Model have been experimentally observed, recently including the Higgs boson.[1][2]

7.1.1 Fermions

Main article: Fermion

Fermions are one of the two fundamental classes of particles, the other being bosons. Fermion particles are described by Fermi–Dirac statistics and have quantum numbers described by the Pauli exclusion principle. They include the quarks and leptons, as well as any composite particles consisting of an odd number of these, such as all baryons and many atoms and nuclei.

Fermions have half-integer spin; for all known elementary fermions this is $\frac{1}{2}$. All known fermions, except neutrinos, are also Dirac fermions; that is, each known fermion has its own distinct antiparticle. It is not known whether the neutrino is a Dirac fermion or a Majorana fermion.[3] Fermions are the basic building blocks of all matter. They are classified according to whether they interact via the color force or not. In the Standard Model, there are 12 types of elementary fermions: six quarks and six leptons.

Quarks

Main article: Quark

Quarks are the fundamental constituents of hadrons and interact via the strong interaction. Quarks are the only known carriers of fractional charge, but because they combine in groups of three (baryons) or in groups of two with antiquarks (mesons), only integer charge is observed in nature. Their respective antiparticles are the antiquarks, which are identical except for the fact that they carry the opposite electric charge (for example the up quark carries charge $+\frac{2}{3}$, while the up antiquark carries charge $-\frac{2}{3}$), color charge, and baryon number. There are six flavors of quarks; the three positively charged quarks are called "up-type quarks" and the three negatively charged quarks are called "down-type quarks".

Leptons

Main article: Leptons

Leptons do not interact via the strong interaction. Their respective antiparticles are the antileptons which are identical, except for the fact that they carry the opposite electric charge and lepton number. The antiparticle of an electron is an antielectron, which is nearly always called a "positron" for historical reasons. There are six leptons in total; the three charged leptons are called "electron-like leptons", while the neutral leptons are called "neutrinos". Neutrinos are known to oscillate, so that neutrinos of definite flavor do not have definite mass, rather they exist in a superposition of mass eigenstates. The hypothetical heavy right-handed neutrino, called a "sterile neutrino", has been left off the list.

7.1.2   Bosons

Main article: Boson

Bosons are one of the two fundamental classes of particles, the other being fermions. Bosons are characterized by Bose–Einstein statistics and all have integer spins. Bosons may be either elementary, like photons and gluons, or composite, like mesons.

The fundamental forces of nature are mediated by gauge bosons, and mass is believed to be created by the Higgs field. According to the Standard Model the elementary bosons are:

The graviton is added to the list although it is not predicted by the Standard Model, but by other theories in the framework of quantum field theory. Furthermore, gravity is non-renormalizable. There are a total of eight independent gluons. The Higgs boson is postulated by the electroweak theory primarily to explain the origin of particle masses. In a process known as the "Higgs mechanism", the Higgs boson and the other gauge bosons in the Standard Model acquire mass via spontaneous symmetry breaking of the SU(2) gauge symmetry. The Minimal Supersymmetric Standard Model (MSSM) predicts several Higgs bosons. A new particle expected to be the Higgs boson was observed at the CERN/LHC on March 14, 2013, around the energy of 126.5GeV with an accuracy of close to five sigma (99.9999%, which is accepted as definitive). The Higgs mechanism giving mass to other particles has not been observed yet.

7.1.3   Hypothetical particles

Supersymmetric theories predict the existence of more particles, none of which have been confirmed experimentally as of 2014:

Note: just as the photon, Z boson and $W^\pm$ bosons are superpositions of the B^0, W^0, W^1, and W^2 fields – the photino, zino, and wino$^\pm$ are superpositions of the bino^0, wino^0, wino^1, and wino^2 by definition.

No matter if one uses the original gauginos or this superpositions as a basis, the only predicted physical particles are neutralinos and charginos as a superposition of them together with the Higgsinos.

Other theories predict the existence of additional bosons:

Mirror particles are predicted by theories that restore parity symmetry.

"Magnetic monopole" is a generic name for particles with non-zero magnetic charge. They are predicted by some GUTs.

"Tachyon" is a generic name for hypothetical particles that travel faster than the speed of light and have an imaginary rest mass.

Preons were suggested as subparticles of quarks and leptons, but modern collider experiments have all but ruled out their existence.

Kaluza–Klein towers of particles are predicted by some models of extra dimensions. The extra-dimensional momentum is manifested as extra mass in four-dimensional spacetime.

7.2 Composite particles

7.2.1 Hadrons

Main article: Hadron

Hadrons are defined as strongly interacting composite particles. Hadrons are either:

- Composite fermions, in which case they are called baryons.

- Composite bosons, in which case they are called mesons.

Quark models, first proposed in 1964 independently by Murray Gell-Mann and George Zweig (who called quarks "aces"), describe the known hadrons as composed of valence quarks and/or antiquarks, tightly bound by the color force, which is mediated by gluons. A "sea" of virtual quark-antiquark pairs is also present in each hadron.

Baryons

See also: List of baryons

Ordinary baryons (composite fermions) contain three valence quarks or three valence antiquarks each.

- Nucleons are the fermionic constituents of normal atomic nuclei:
  - Protons, composed of two up and one down quark (uud)
  - Neutrons, composed of two down and one up quark (ddu)
- Hyperons, such as the Λ, Σ, Ξ, and Ω particles, which contain one or more strange quarks, are short-lived and heavier than nucleons. Although not normally present in atomic nuclei, they can appear in short-lived hypernuclei.
- A number of charmed and bottom baryons have also been observed.

Some hints at the existence of exotic baryons have been found recently; however, negative results have also been reported. Their existence is uncertain.

- Pentaquarks consist of four valence quarks and one valence antiquark.

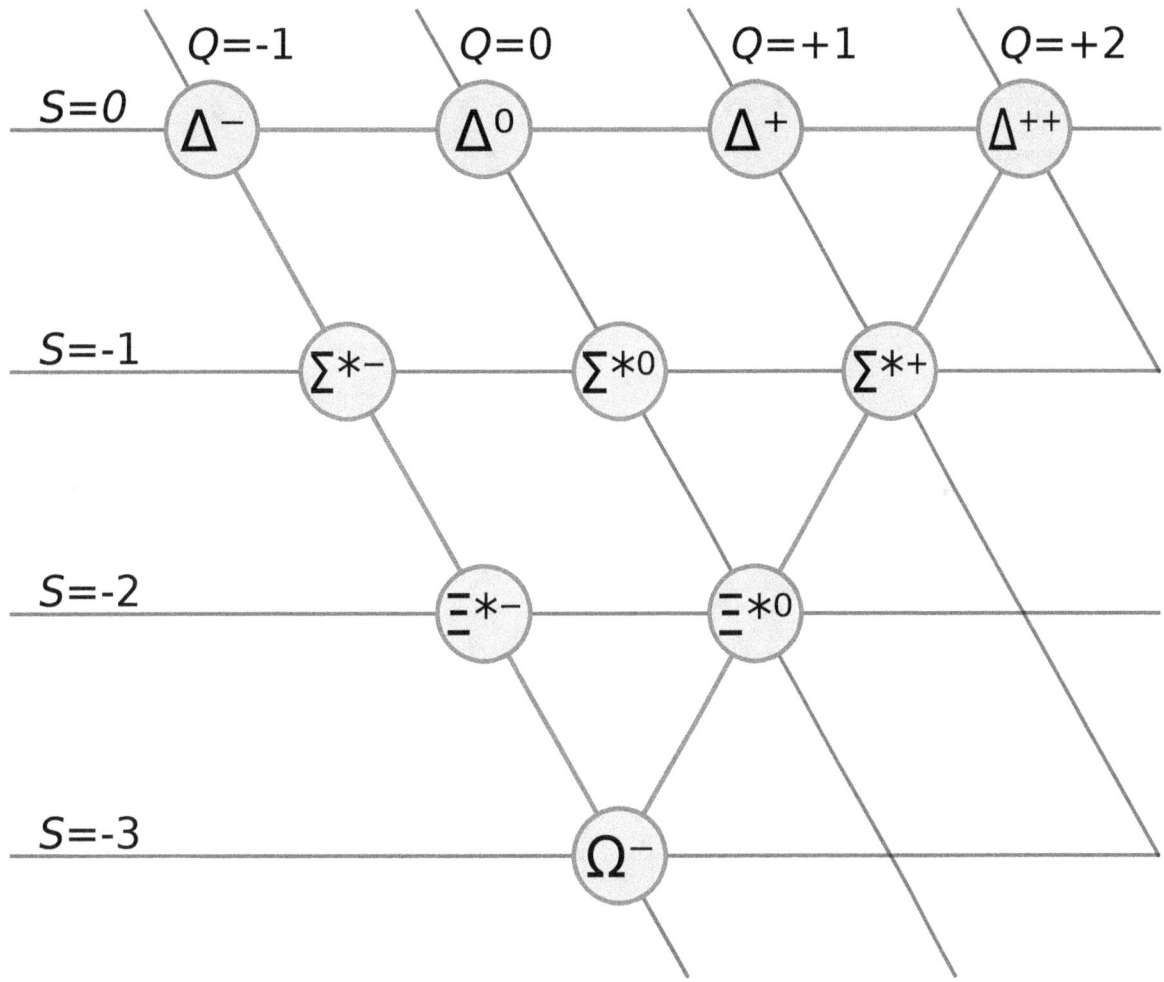

A combination of three u, d or s-quarks with a total spin of $^3/_2$ form the so-called "baryon decuplet".

Mesons

See also: List of mesons

Ordinary mesons are made up of a valence quark and a valence antiquark. Because mesons have spin of 0 or 1 and are not themselves elementary particles, they are "composite" bosons. Examples of mesons include the pion, kaon, and the J/ψ. In quantum hydrodynamic models, mesons mediate the residual strong force between nucleons.

At one time or another, positive signatures have been reported for all of the following exotic mesons but their existences have yet to be confirmed.

- A tetraquark consists of two valence quarks and two valence antiquarks;

- A glueball is a bound state of gluons with no valence quarks;

- Hybrid mesons consist of one or more valence quark-antiquark pairs and one or more real gluons.

7.2.2 Atomic nuclei

Atomic nuclei consist of protons and neutrons. Each type of nucleus contains a specific number of protons and a specific

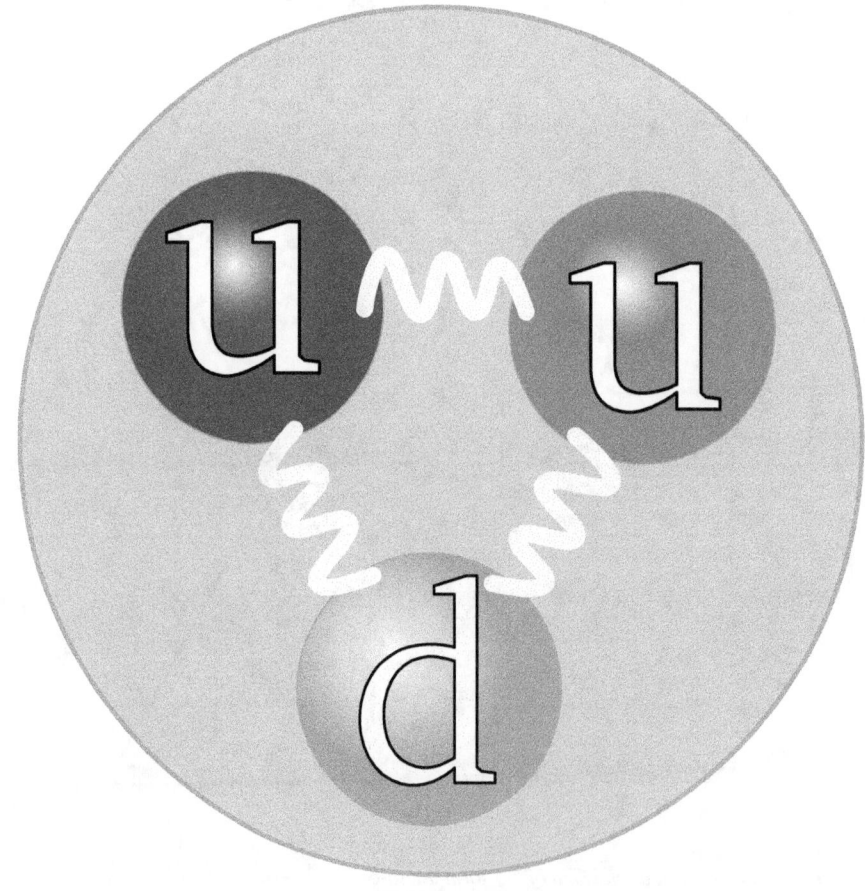

Proton quark structure: 2 up quarks and 1 down quark. The gluon tubes or flux tubes are now known to be Y shaped.

number of neutrons, and is called a "nuclide" or "isotope". Nuclear reactions can change one nuclide into another. See table of nuclides for a complete list of isotopes.

7.2.3 Atoms

Atoms are the smallest neutral particles into which matter can be divided by chemical reactions. An atom consists of a small, heavy nucleus surrounded by a relatively large, light cloud of electrons. Each type of atom corresponds to a specific chemical element. To date, 118 elements have been discovered, while only the elements 1-112,114, and 116 have received official names.

The atomic nucleus consists of protons and neutrons. Protons and neutrons are, in turn, made of quarks.

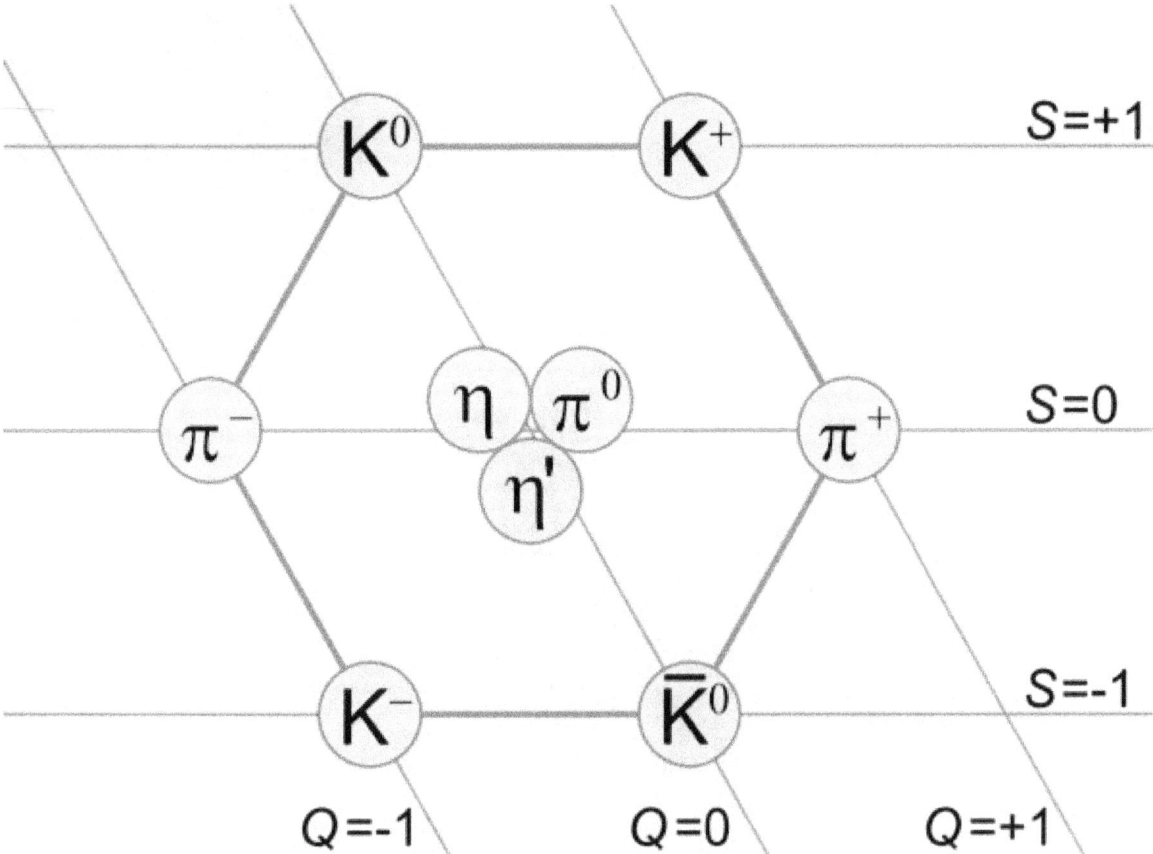

Mesons of spin 0 form a nonet

7.2.4 Molecules

Molecules are the smallest particles into which a non-elemental substance can be divided while maintaining the physical properties of the substance. Each type of molecule corresponds to a specific chemical compound. Molecules are a composite of two or more atoms. See list of compounds for a list of molecules.

7.3   Condensed matter

The field equations of condensed matter physics are remarkably similar to those of high energy particle physics. As a result, much of the theory of particle physics applies to condensed matter physics as well; in particular, there are a selection of field excitations, called quasi-particles, that can be created and explored. These include:

- Phonons are vibrational modes in a crystal lattice.

- Excitons are bound states of an electron and a hole.

- Plasmons are coherent excitations of a plasma.

- Polaritons are mixtures of photons with other quasi-particles.

- Polarons are moving, charged (quasi-) particles that are surrounded by ions in a material.

- Magnons are coherent excitations of electron spins in a material.

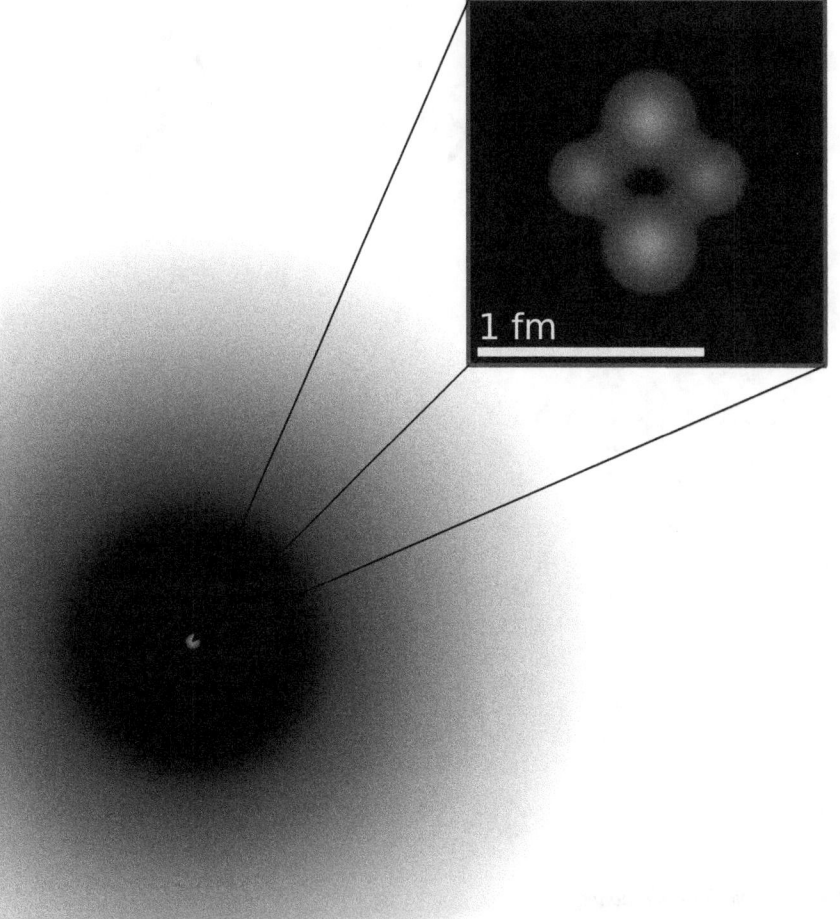

1 Å = 100,000 fm

A semi-accurate depiction of the helium atom. In the nucleus, the protons are in red and neutrons are in purple. In reality, the nucleus is also spherically symmetrical.

7.4 Other

- An anyon is a generalization of fermion and boson in two-dimensional systems like sheets of graphene that obeys braid statistics.

- A plekton is a theoretical kind of particle discussed as a generalization of the braid statistics of the anyon to dimension > 2.

- A WIMP (weakly interacting massive particle) is any one of a number of particles that might explain dark matter (such as the neutralino or the axion).

- The pomeron, used to explain the elastic scattering of hadrons and the location of Regge poles in Regge theory.

- The skyrmion, a topological solution of the pion field, used to model the low-energy properties of the nucleon, such as the axial vector current coupling and the mass.

- A genon is a particle existing in a closed timelike world line where spacetime is curled as in a Frank Tipler or Ronald Mallett time machine.

- A goldstone boson is a massless excitation of a field that has been spontaneously broken. The pions are quasi-goldstone bosons (quasi- because they are not exactly massless) of the broken chiral isospin symmetry of quantum chromodynamics.

- A goldstino is a goldstone fermion produced by the spontaneous breaking of supersymmetry.

- An instanton is a field configuration which is a local minimum of the Euclidean action. Instantons are used in nonperturbative calculations of tunneling rates.

- A dyon is a hypothetical particle with both electric and magnetic charges.

- A geon is an electromagnetic or gravitational wave which is held together in a confined region by the gravitational attraction of its own field energy.

- An inflaton is the generic name for an unidentified scalar particle responsible for the cosmic inflation.

- A spurion is the name given to a "particle" inserted mathematically into an isospin-violating decay in order to analyze it as though it conserved isospin.

- What is called "true muonium", a bound state of a muon and an antimuon, is a theoretical exotic atom which has never been observed.

7.5   Classification by speed

- A tardyon or bradyon travels slower than light and has a non-zero rest mass.

- A luxon travels at the speed of light and has no rest mass.

- A tachyon (mentioned above) is a hypothetical particle that travels faster than the speed of light and has an imaginary rest mass.

7.6   See also

- Acceleron

- List of baryons

- List of compounds for a list of molecules.

- List of fictional elements, materials, isotopes and atomic particles

- List of mesons

- Periodic table for an overview of atoms.

- Standard Model for the current theory of these particles.

- Table of nuclides

- Timeline of particle discoveries

7.7 References

[1] Observation of a new boson at a mass of 125 GeV with the CMS experiment at the LHC (2013). *arXiv:1207.7235*.

[2] Observation of a new particle in the search for the Standard Model Higgs boson with the ATLAS detector at the LHC (2012). *arXiv:1207.7214*.

[3] B. Kayser, *Two Questions About Neutrinos*, arXiv:1012.4469v1 [hep-ph] (2010).

[4] R. Maartens (2004). *Brane-World Gravity* (PDF). *Living Reviews in Relativity* **7**. p. 7. Also available in web format at http://www.livingreviews.org/lrr-2004-7.

- C. Amsler *et al.* (Particle Data Group) (2008). "Review of Particle Physics". *Physics Letters B* **667** (1–5): 1. Bibcode:2008PhLB..667....1P. doi:10.1016/j.physletb.2008.07.018. *(All information on this list, and more, can be found in the extensive, biannually-updated review by the Particle Data Group)*

7.8 Text and image sources, contributors, and licenses

7.8.1 Text

- **Positron** *Source:* https://en.wikipedia.org/wiki/Positron?oldid=679431995 *Contributors:* Bryan Derksen, Andre Engels, Hhanke, Peterlin~enwiki, Patrick, Looxix~enwiki, Hashar, Stismail, Pstudier, BenRG, Donarreiskoffer, Robbot, Merovingian, LGagnon, David Gerard, Decumanus, Giftlite, Jmnbpt, Xerxes314, Utcursch, Xmnemonic, Knutux, Icairns, Tumbarumba, Mike Rosoft, Vsmith, Jpk, Murtasa, Bender235, El C, Asierra~enwiki, La goutte de pluie, Tra, Atlant, Spangineer, Wtmitchell, SidP, RogerBarnett, Gene Nygaard, WojciechSwiderski~enwiki, UTSRelativity, Richard Arthur Norton (1958-), MartinSpacek, Jannex, Ma Baker, Robert K S, Palica, MassGalactusUniversum, V8rik, BD2412, Canderson7, Seraphimblade, Mike Peel, FlaBot, DannyWilde, Kolbasz, Fresheneesz, Tardis, Wrightbus, Chobot, YurikBot, Bambaiah, Jimp, Arado, Chuck Carroll, Bergsten, Damato, Hellbus, Astriks, Salsb, MidnightWolf, Vanished user kjdioejh329io3rksdkj, Complainer, Robertvan1, Howcheng, SCZenz, Zwobot, Scottfisher, Lt-wiki-bot, Arthur Rubin, Terbospeed, Vicarious, GrinBot~enwiki, SmackBot, Unyoyega, Edgar181, KingRaptor, Skizzik, Kmarinas86, Complexica, DHN-bot~enwiki, Sbharris, Rodri316, JorisvS, Mr. Vernon, Rock4arolla, MTSbot~enwiki, Iridescent, Michaelbusch, Darth Sader, CapitalR, CmdrObot, Eric, Cofax48, Yzphub, HalJor, Cydebot, Nick Y., Bvcrist, Yolocavo, HPaul, Quibik, Thijs!bot, Gamer007, Headbomb, Rlupsa, Davidhorman, Shadow Blaziken, Griba2010, Escarbot, Poshzombie, Harrylentil, JAnDbot, CosineKitty, Rob Mahurin, MegX, Recurring dreams, Mother.earth, Jetterman, Maliz, Gwern, Geboy, Boddey, Rustyfence, CliffC, Rigmahroll, CommonsDelinker, HEL, Solarswordsman, AntiSpamBot, Y2H, Sheliak, TreasuryTag, Philip Trueman, TXiKiBoT, DavidRThomas, Corticopia, Agradada, Corvus cornix, Lerdthenerd, Gabriel Vidal, Seraphita~enwiki, AlleborgoBot, SieBot, Winchelsea, Triwbe, RadicalOne, Flyer22, WingkeeLEE, BenoniBot~enwiki, Drgarden, ClueBot, Fyyer, Gabriel Vidal Álvarez, Piledhigheranddeeper, Maxtian, Djr32, Tyler, EncyclopediaUpdaticus, DumZiBoT, AgnosticPreachersKid, TheRealVolucrix, SkyLined, Addbot, DOI bot, AkhtaBot, CanadianLinuxUser, Jim10701, AgadaUrbanit, Positroni, Tide rolls, Luckas-bot, Yobot, Allowgolf~enwiki, Götz, Jim1138, Taupositron, Materialscientist, Citation bot, Yathimc, Jakouso, Stephen.G.McAteer, FrescoBot, Paine Ellsworth, Dscraggs, Citation bot 1, I dream of horses, Tom.Reding, Wdanbae, RjwilmsiBot, Samdacruel, EmausBot, WikitanvirBot, Harddk, H3llBot, Quondum, I kabir, MG-Weatherman08, Carmichael, Teapeat, ClueBot NG, Snotbot, Helpful Pixie Bot, Ieditpagesincorrectly, Bibcode Bot, Slaughter182, Sergeant Cribb, Drizzt182, Dexbot, Crocgandhi, Bigdumpy, Planetguy2345, KasparBot, Kobiej100, People73, Person420 and Anonymous: 148

- **Antiparticle** *Source:* https://en.wikipedia.org/wiki/Antiparticle?oldid=680600374 *Contributors:* AxelBoldt, CYD, Mav, Bryan Derksen, Andre Engels, Josh Grosse, Stevertigo, Mrwojo, Patrick, RTC, Paddu, CesarB, Nikai, Nikola Smolenski, Charles Matthews, The Anomebot, Wik, Omegatron, Bevo, Altenmann, Merovingian, Intangir, Wikibot, Martinwguy, Giftlite, Bogdanb, Harp, BenFrantzDale, Herbee, Spencer195, Fleminra, Jason Quinn, Zeimusu, Mako098765, Karol Langner, Mike Rosoft, Helohe, Rich Farmbrough, Guanabot, Pjacobi, Guanabot2, Mr. Billion, Joanjoc~enwiki, Kghose, Cmdrjameson, Giraffedata, Matt McIrvin, HasharBot~enwiki, Pediddle, Deror avi, Woohookitty, Mindmatrix, Wdyoung, GregorB, SeventyThree, Justin Ormont, Palica, Marudubshinki, Tevatron~enwiki, Rjwilmsi, Ae77, MZMcBride, KaiMartin, FlaBot, Krackpipe, Commander Nemet, Roboto de Ajvol, YurikBot, Borgx, Bambaiah, Zhaladshar, Spike Wilbury, Bota47, Terbospeed, Mkossick, Tim314, 四葉亭robot, SmackBot, FocalPoint, Alsandro, Srnec, Dauto, Octahedron80, Drphilharmonic, Marcus Brute, Vinaiwbot~enwiki, Jake-helliwell, Grumpyyoungman01, Newone, Mellery, Van helsing, Tim1988, Myasuda, Gogo Dodo, Goldencako, Thijs!bot, Headbomb, Tyco.skinner, JAnDbot, Steveprutz, Ferritecore, Jpod2, Singularity, Dbiel, TomasBat, Eternalmatt, Joshmt, DorganBot, Cuckooman4, VolkovBot, TXiKiBoT, Red Act, Anonymous Dissident, AlleborgoBot, SieBot, Likebox, RadicalOne, Flyer22, KoenDelaere, Thomega, RW Marloe, BrightRoundCircle, Davidmosen, Jacob.jose, Anyeverybody, ClueBot, Diagramma Della Verita, Alexbot, Eeekster, Rishi.bedi, SilvonenBot, NellieBly, Lilaspastia, SkyLined, AkhtaBot, CarsracBot, Lightbot, Legobot, Luckas-bot, Yobot, Planlips, Csmallw, AnomieBOT, Citation bot, Vuerqex, ArthurBot, Xqbot, Omnipaedista, RibotBOT, Muhwang, EmausBot, John of Reading, L Kensington, Benazhack, ClueBot NG, Geekingreen, Mesoderm, Bibcode Bot, B wik, Mark Arsten, Rm1271, Penguinstorm300, Robotsheepboy, YFdyh-bot, 77Mike77, संजीव कुमार, Dert567, Monkbot, Nazo!nin, KasparBot and Anonymous: 100

- **Positron emission** *Source:* https://en.wikipedia.org/wiki/Positron_emission?oldid=669450942 *Contributors:* AxelBoldt, Stone, Kbk, Pstudier, BenRG, Donarreiskoffer, Henrygb, Herbee, Mike40033, Icairns, Vsmith, Sunborn, StradivariusTV, Benbest, Strait, Whitlock, Carrionluggage, Chobot, Sliggy, Newagelink, Geoffrey.landis, Mako098765, Sbharris, Peterwhy, Dale101usa, Tuttt, Comrade42, Harej bot, MrFizyx, Headbomb, Stannered, Rob Mahurin, VoABot II, Leyo, Relativeculo, Trumpet marietta 45750, Sheliak, TXiKiBoT, Hqb, Judge Nutmeg, Pamputt, Denisarona, ClueBot, Mild Bill Hiccup, Jaburgess10, SkyLined, Addbot, The CIS, Spacy73, Hartz, Yobot, The High Fin Sperm Whale, Citation bot, LilHelpa, FrescoBot, LucienBOT, Minivip, Double sharp, Jojofunny123, Yangosplat222, Rami radwan, JSquish, Makecat, Geo7777, Bibcode Bot, BG19bot, Zedshort, Arcandam, Makecat-bot, SongCloud, KasparBot and Anonymous: 32

- **Positron emission tomography** *Source:* https://en.wikipedia.org/wiki/Positron_emission_tomography?oldid=682353202 *Contributors:* AxelBoldt, Magnus Manske, The Anome, Malcolm Farmer, Alex.tan, Andre Engels, Fnielsen, Rmhermen, Rsabbatini, Patrick, RTC, Michael Hardy, Vaughan, David Martland, Karada, Andrel, Andrewa, Julesd, Salsa Shark, Jiang, Rl, Schneelocke, RodC, Wikid, Doradus, Slakhan, J D, Oaktree b, Phil Boswell, Robbot, JamesMLane, DocWatson42, Dratman, Wolfgangamadeus, Khalid hassani, Bobblewik, Delta G, SoWhy, Ilgiz, Sayeth, Neutrality, Deglr6328, Discospinster, MAlvis, Bender235, Flapdragon, Zaslav, Chtito, Wisdom89, Arcadian, Kjkolb, Jumbuck, Larham, Alansohn, Atlant, Improv, Wouterstomp, Axl, Wtmitchell, Cburnett, Danhash, TenOfAllTrades, Fivepints, Zereshk, Oliphaunt, Ropcat, CharlesC, Graham87, Rjwilmsi, SeanMack, Lcolson, FlaBot, RobertG, Salvadorjo~enwiki, Chobot, YurikBot, RobotE, Damato, Gaius Cornelius, NawlinWiki, A314268, Jeremy Butler, Andrew73, MacsBug, SmackBot, LostIntel, Mathewbrowne, Pgk, Chetanr, Anastrophe, Benjaminevans82, Kazkaskazkasako, Tito4000, Fizzy, Kostmo, Sbharris, Tekhnofiend, Chlewbot, TonySt, Makemi, AdeMiami, Drphilharmonic, DMacks, Lambiam, John, Shlomke, Ben Moore, Astuishin, Slakr, ChumpusRex, Hu12, Ccohalan, UncleDouggie, Captainj, Chirality, JForget, CmdrObot, Nunquam Dormio, Dgw, Cydebot, Gogo Dodo, Was a bee, Anthonyhcole, Naffer, Epbr123, Kubanczyk, Purple Paint, Headbomb, Electron9, EdJohnston, OrenBochman, AntiVandalBot, Shirt58, JEH, Dougher, JAnDbot, Kaobear, Db099221, Yeti7, Dricherby, .anacondabot, Acroterion, Magioladitis, VoABot II, Ojh2, BlueEvo2, SineWave, Mark PEA, Alex Spade, Spellmaster, JdeJ, Crampedson, DGG, Duedilly, Pernambuco, Jascii, J.delanoy, NikNakk, Boghog, Yhseo, MrBell, Sjayanthi, Kygkim, Bdekker, Mrs.meganmmc, Sduncan53, Bfong, Mikael Häggström, Junior Brian, Gaussgauss, Sleepeeg3, Tevonic, Bonadea, Hehkuviini, Philip Trueman, TXiKiBoT, Antoni Barau, Ask123, Anna Lincoln, Stevesg, Jk91185, SnowflakeHolocaust, Doc James, Neparis, SieBot, Da Joe, Caltas, Lucasbfrbot, Flyer22, King Spadina, Permacultura, Harsh Stone White, Drgarden, Sfan00 IMG, ClueBot, Unbuttered Parsnip, CounterVandalismBot, Phdplayahatadegree, Frederickhoyles, Bertrus, Johnsaunderson, T.j.chryssikos, Crud302, Versus22, DumZiBoT, EdChem, Markssss, CallipygianSchoolGirl, Mco44, Good

7.8.2 Images

7.8.3 Content license

www.ingramcontent.com/pod-product-compliance
Lightning Source LLC
Chambersburg PA
CBHW080559180526

45168CB00007B/2713